I0484908

Evolution Collides with Science and Reason

By
Lewis H. Clementson, J.D.

Printed in the United States of America.

The author was graduated by the University of Virginia with a BA in biology, and with a Juris Doctor by George Mason University School of Law. He has practiced law for over 30 years, written six books and served as a missionary to France and Russia.
Several of his books are available on internet sale sites and lewisclementsonministries.org

Table of Contents

Chapter 1
It Happened One Night

It was 1981 and there was a meeting at the American Museum of Natural History in New York City. It was certainly not a big deal to most of the city's eight million residents. And, although their guest speaker was the very well-known senior paleontologist from the British Museum of Natural History in London, no one present could have possibly foreseen the firestorm that would erupt that night.

The renowned Dr. Colin Patterson got up to speak and immediately asked his distinguished audience:

> "Can you tell me anything you know about evolution, any one thing, any one thing that is true?
>
> I tried that question on the geology staff at the Field Museum of Natural History and the only answer I got was silence. I tried it on the members of the Evolutionary Morphology Seminar in the University of Chicago, a very prestigious body of evolutionists, and all I got there was silence for a long time and eventually one person said, 'I do know one thing. It ought not to be taught in high school.'"

The crowd in New York was not to be any different. His question hung out there like a red cape in front of a bull, but still no one spoke up. There was only silence from the entire audience, and this audience consisted almost entirely of men and women who would have jumped at the opportunity to engage in debate with this famous and important man in front of so many distinguished colleagues and, at the same time, defend their beloved theory. Nevertheless, no one raised a hand. No one took advantage of the career opportunity of a lifetime, because no one had anything to say.

Again Dr. Patterson repeated the question. Again the only response was the noise from people squirming in their seats.

Dr. Patterson looked around the entire room, an open invitation written all over his face. Finally Dr. Patterson said categorically,

> "The absence of answers seems to suggest that it is true, evolution does not convey any knowledge, or if so, I haven't yet heard of it."

Still, no one spoke up. No one in this crowd had any evidence or knew of anything true about evolution. And if there really was anything true about evolution, this crowd would have known. But no one jumped up and shouted his outrage at such a ridiculous question. No one indicated that the question was even an insult. In fact, no one seemed at all surprised by the question. Instead, there was only dead silence, and what an amazing silence it was. A distinguished group of devoted evolutionists was admitting to the senior paleontologist from the British Museum of Natural History in London, and to themselves, that there was nothing about evolution, not any one thing that was true.

But what they did not know was that there was a spy in the audience. A creationist had sneaked in and heard it all. He had not only heard the entire speech, he had a tape recorder hidden on him and recorded the whole thing.[1]

The firestorm was lit.

Creationists immediately published the speech and evolutionists descended upon poor Dr. Patterson and did everything but put a gun to his head to force him to retract his conclusions. He issued several different complicated explanations, but none of them would qualify as a retraction. He only accentuated his position that evolution had a gaping hole in its theory. No one could show "how" evolution could happen. Ever since, evolutionists have desperately tried to explain away his remarks, but really what they should have done is answer the question, not attack the speaker. But they have not, because they cannot.

Nevertheless, evolution remains the predominant worldwide belief about the origin of all life on earth. And evolution is not just some obscure scientific theorem that is irrelevant to most non-scientists. It is extremely relevant to every one of us. Evolutionists are telling me that I, a man with all the same self-awareness that everyone else has, am not

---

[1] The full text of the transcript of his talk is copyrighted by Stephen E. Jones but can be viewed online at
http://members.iinet.net.au/~sejones/pattamnh.html
If you can't access with this, try a web search of "Stephen E. Jones: Colin Patterson's address at the American Museum of Natural History".

created in the image of God at all but am a mere accident of nature and only one accidental step up from an ape at that. This is getting as personal as you can get.

The empty space at the end of each chapter seemed like a good place to put a few select photos of nature's wonder, just for your viewing pleasure.

Black Panther

Chapter 2
At the University of Virginia

At age eighteen, I entered the University of Virginia and majored in biology because I wanted to be a marine biologist. I took every course required by the department for that major. I took courses in biology, general chemistry, physics, general physiology, organic chemistry, genetics, embryology and anatomy, biology of bacteria, comprehensive neurophysiology, human evolution and general evolution, all of which amounted to roughly one half of my total four years at UVA. It was a rigorous curriculum, but I loved science. I loved how the discipline revealed the mysteries of our world and nature. I loved the marvelous cycles and systems that operate on their own and in perfect harmony with each other, and the wonders of physics and of chemistry. I marveled that a nucleus with six protons forms carbon, one of the hardest elements on earth, and a nucleus with seven protons forms nitrogen, a gas.

Fascinating stuff, but toward the end of my college career, I visited what was the largest marine biology facility in America at the time, which happened to be in my home state of Virginia, and the whole place was smaller than our house.[2] And our house was not large. So I considered the job prospects,

---

[2] Virginia Institute of Marine Science, located in Gloucester Point, VA. Now it also boasts the *School of Marine Science (SMS) at VIMS* which is the graduate school in marine science for the College of William & Mary.

began looking elsewhere, and ended up a lawyer. But my love for science has stayed with me.

But there was one thing about my studies at UVA that bothered me then and still bothers me now. And that is the special treatment afforded evolution in the scientific community.

I still remember how the first biology class started. The class was taught by the chairman of the department, and he started our college career off by saying in a most threatening tone, "If there is anyone here who *questions* the *reality* of evolution, you may see me in my office after class." That ended that. No debate would be tolerated.

The biology textbooks were full of references to evolution and as I already said, two classes devoted exclusively to evolution were required for the major. It was shoved down our throats, and by comparison with the other subjects, without much to support it. All of our other textbooks were about three inches thick, and each page was crammed with so many facts that I often would highlight the whole page. This was true with every single field of science that I studied. In addition, for each of the main subjects, there were numerous volumes of outside reading if you wanted more. The amount of information was absolutely staggering.

And then there was evolution. The textbook was a half inch thick, if that, and most of that meager half inch was about Charles Darwin's voyage on the ship *Beagle* to the Galapagos Islands and a presentation of the theory. When I was reading the book, I groped for something, anything, to underline that might be a test question. I read whole chapters and realized that I had not underlined anything and had to go back to look for something.

The scientific 'evidence' was extremely limited. It included a study of the peppered moth in England and how the dark variety prospered in the soot covered areas and lighter colored ones prospered where there was no soot. But they were all just variations of the same species, like breeds of dogs are all dogs. It also covered a snail species in a valley on an island and how it varied with its environment. But again, they were all the same species. We were accustomed to studying science with scientific facts and scientific tests, and lots and lots of data. There was obviously no real evidence for this theory, in our college textbook. The comparison with all the other subjects was striking. It had somehow acquired a special exemption from the proofs and standards imposed on all other areas of science.

The other problem was the professor's *attitude* about evolution, which was quite defensive. Many questions were posed in the evolution classes that were not answered. Many! And how the professor squirmed. But regardless of the numerous and significant problems, the faculty remained adamant about it and exhibited a defensiveness not present with any other subject.

Another thing that was immediately apparent in the study of science was its complexity. I cannot exaggerate how very complex life is, and how intricately it operates in harmony with the rest of nature, both inside the various life forms and also with other systems on the planet. Yet the biology faculty remained adamant that it all happened purely *by accident.* Something was wrong, obviously very wrong.

Chapter 3
Basic Science

In order to understand one of the main problems with evolution, it will be very helpful if you understand a little bit about nature and science. One of the main reasons that the evolution myth has survived is that most people think that science should be left to the scientists, and that they are not smart enough to understand the science behind it. Neither is true. So, I will attempt a basic science lesson. The material presented here is nothing more than what an average 18 year old is asked to learn in a very short time. If you will just read this little bit of science, you will have a whole new appreciation for the complexity of life and also the very issue at hand. So please be patient. There really is a good reason to cover this.

Atoms
All matter of every kind is made up of atoms. Whether it is solid, liquid or gas, it is still made up of atoms. Atoms themselves are made up of sub-atomic particles; mainly electrons, protons and neutrons. The protons and neutrons are tightly bound in the atom's nucleus and the electrons[3] spin in orbit around the nucleus. Electrons have a negative magnetic charge and the protons have a positive magnetic charge. The

---

[3] The diameter of an electron is 1/1000 the diameter of a proton. Its mass in grams can be written with a decimal point followed by 27 zeros and a 9.

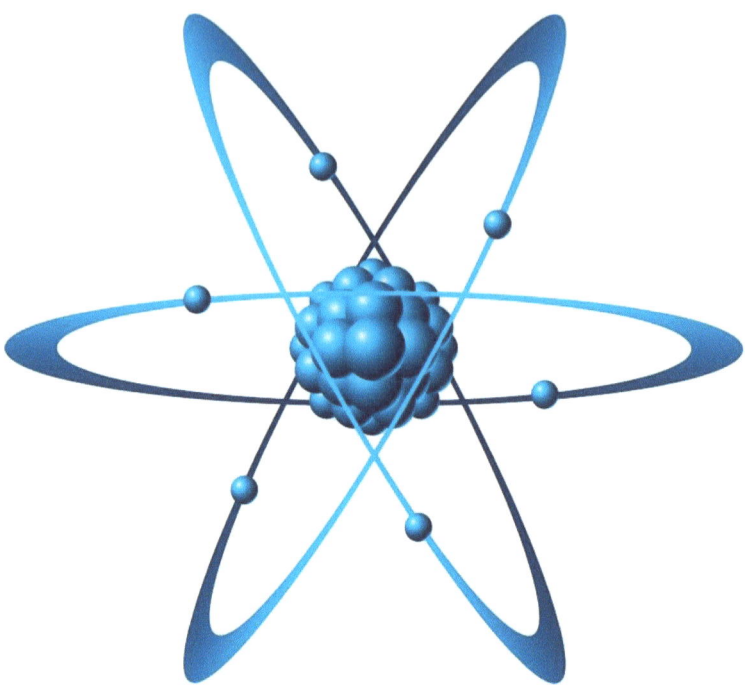

negatively charged electrons are held in orbit by the pull from the positive charge of the protons in the nucleus. For each negatively charged electron in orbit there is a corresponding positively charged proton in the nucleus. There is also one neutron in the nucleus for every proton, and as the name implies, neutrons are neutral and have no magnetic charge. There are 103 different elements in nature, each of which has a different number of protons and a corresponding number of electrons. For instance, carbon has 6 protons and 6 electrons.

Electrons are the force in electricity, lightning and static electricity. When we were children, we rubbed our shoes on the carpet to scoop up electrons and then touched someone to give them a shock of static electricity. An atom can lose an electron and not lose its character as a specific element. However, if an atom loses or gains a proton, it becomes a different element.

Electrons revolve around the atom's nucleus in different levels of orbit, or quantum. Naturally, a higher orbit involves more energy than a lower orbit. If heat is applied to an atom, an electron can jump from a lower orbit to a higher orbit, and if its temperature drops, then the electron will drop to a lower orbit. When it drops to a lower orbit, it emits light energy. This is why a heated metal will emit a red color.[4]

We know from experience that two positively charged things will repel each other. If you have ever tried to push the positive ends of two magnets together, you have seen this force at work. This is also true with subatomic particles. Two or more protons would not normally sit comfortably side by side in a nucleus. They would repel each other. However, all atoms with two or more protons have protons that sit comfortably side by side in the nucleus. The powerful energy that holds multiple positively charged protons together side by side in an atom's nucleus is called nuclear energy. This is a very mysterious aspect of nature, and it was this mystery that captured Albert Einstein's curiosity. He wanted to calculate how much energy it takes to hold protons together. It was apparent that the more protons in the nucleus, the more nuclear energy was needed. Therefore the more mass there is, the more nuclear energy is present in

---

[4] The study of electrons, their quantum, their frequency of vibration and how light and heat are released is called "quantum physics" and is fascinating stuff.

the nucleus. But how much energy is required for each unit of mass? He calculated that it was a number equal to the speed of light squared.[5]

$E=mc^2$ (nuclear energy equals mass times the speed of light squared)

Now consider for a moment that the most elemental part of all matter, the nucleus of an atom, contains that much energy.

Molecules

Next, after atoms, there are molecules. Molecules are a combination of two or more atoms held together by the joining of their electrons. The simplest is the hydrogen molecule, which consists of two hydrogen atoms.

The most basic of molecules in all living matter is DNA, (deoxyribonucleic acid). Every species of plant and animal has a different DNA molecule and each individual within each species also has a variation on that. DNA identification is now quite common.

The astounding complexity of the DNA molecule makes it possible for each species to have an almost limitless number of variations. Yet the DNA molecule is the most basic and elementary building block of all life. In nature, this is about as simple as life is going to get, and yet the complexity of its

---

[5] Light travels at the speed of 186,000 miles per second, or 300 million meters per second. Square that and you get 9 with 16 zeros. So, according to Einstein, for every kilogram (2.2 pounds) of matter, it takes 90,000,000,000,000,000 joules of nuclear energy to hold it together. A joule of energy is one watt of power for one second.

structure could easily consume the contents of a very large book.

Watson and Crick's breakthrough discovery of the DNA molecule in 1953 was that the molecule formed a double spiral helix. [6] They received the Nobel Prize for this. [7]

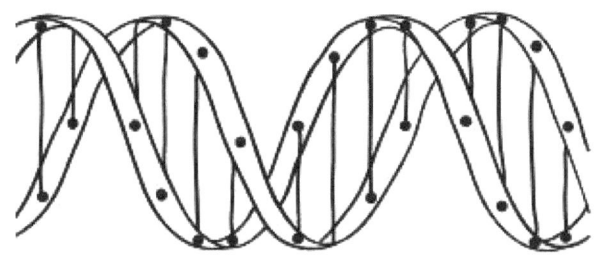

Living organisms consist primarily of protein molecules which are even more complex than DNA. Protein molecules contain between 1,000 and 10,000 atoms. Naturally, each different protein molecule has a different size, sequence of atoms and 3-D shape. They also have a chain-like shape.

Shortly after Watson and Crick's discovery of the DNA *molecule*, Frederick Sanger discovered the

---

[6] Each strand in the double spiral consists of a sugar-phosphate backbone and numerous base atoms, each attached in pairs. The four bases that make up the stair-like pattern in the spiral staircase are adenine, thymine, cytosine and guanine.

[7] In 1962 James Watson (b. 1928), Francis Crick (1916–2004), and Maurice Wilkins (1916–2004) jointly received the Nobel Prize in physiology or medicine for their 1953 determination of the structure of deoxyribonucleic acid (DNA).

sequence of amino acid molecules in an entire *protein*, insulin.[8] This breakthrough took him ten years of meticulous work.

Genes

After molecules, come genes, which are the basic unit of genetics. Genes are tiny particles containing all the information necessary for the growth of each species and every individual in that species. Even the simplest species requires thousands of genes.

Every cell in your body contains all of your genes. It doesn't matter whether the cell is in your nose or your toes; every cell in your body has the same nucleus with the same genes. Genes have been referred to as the "blueprint" or "code of life" for each individual. Every trait, such as eye color and height, are determined by a certain gene or genes. Many biologists have devoted their lives to the study of genes and they know what a gene can do, but no one has yet to figure out how it can do it.

---

7. Frederick Sanger, 1918-2013, a Quaker, proved that proteins were ordered molecules and the genes and DNA that make these proteins have an order or sequence as well. Sanger won his first Nobel Prize for Chemistry in 1958 for his work on the structure of protein.
Sanger won a second Nobel Prize for Chemistry in 1980 sharing it with Walter Gilbert, for their contributions concerning the determination of base sequences in nucleic acids.

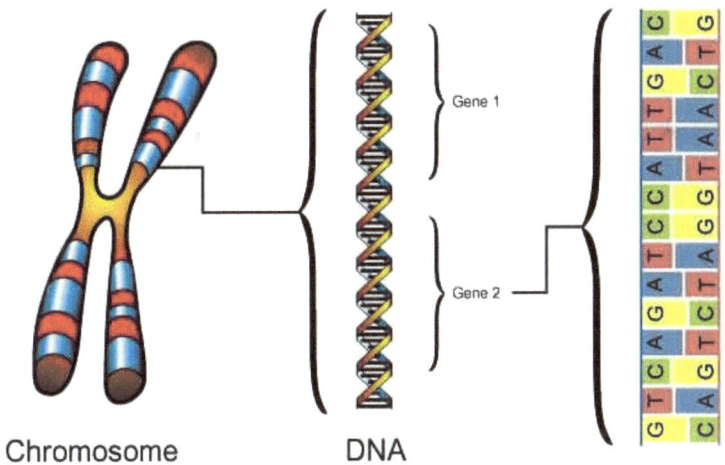

Chromosome   DNA

Genes

Chromosomes

All genes are attached to chromosomes. Each chromosome contains thousands of genes and looks like a furry rope. They come in pairs, lying side by side and attached in the middle, so they take the shape of an "X". Humans have 23 pairs, or 46 chromosomes. Cats have 38, dogs 78, and horses 64.

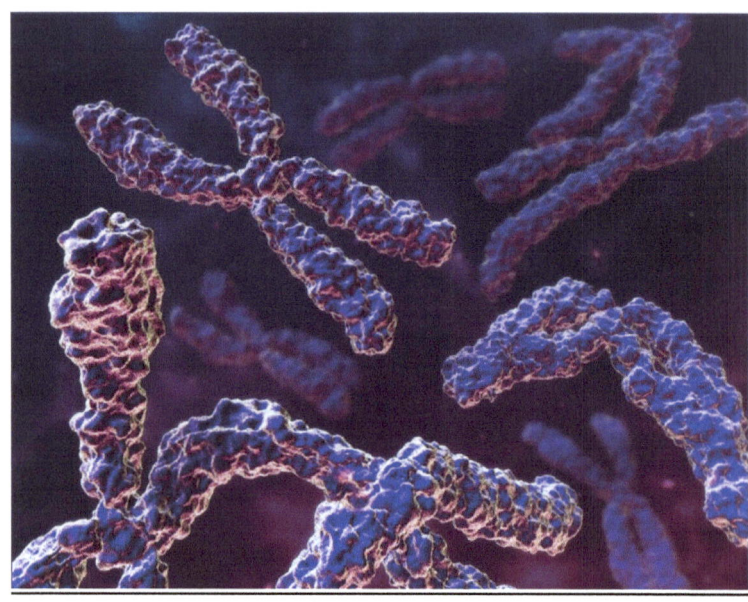

Chromosomes

Cells

In his book <u>The Origin of Species</u>, Charles Darwin constantly referred to the hypothetical "simple single cell". He claimed that it came into existence spontaneously and that all life evolved from it. At that time, atomic physics, microbiology and cellular biology were nonexistent, and the true complexity of even the simplest cell was not fully appreciated. Nothing was known about atoms, molecules, amino acids, genes or chromosomes. But now, a little bit is known, and with it, the knowledge that there is no such thing as a "simple single cell". It simply doesn't exist. There is not even such a thing as a simple atom, molecule, gene or chromosome, much less an entire cell. A bacteria cell is the simplest organism capable of independent life, and a leading authority on this subject, Phillip Johnson, has characterized a bacteria cell as "a masterpiece of miniaturized complexity which makes a spaceship seem rather low-tech."[9]

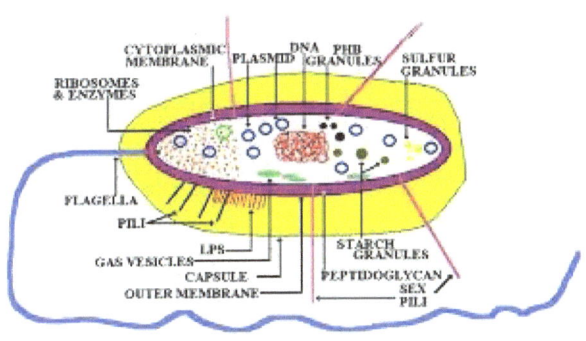

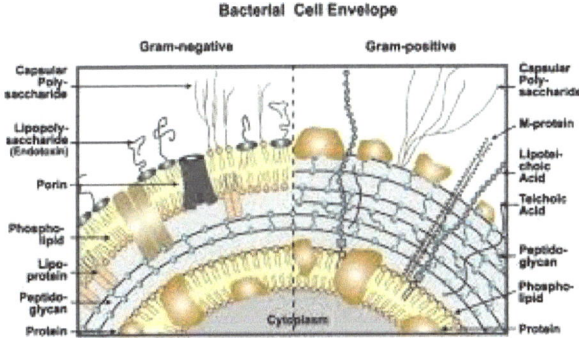

Mitosis

All plants and animals grow and replenish themselves by the dividing of its cells into two identical daughter cells. As the reader now knows, all cells contain a nucleus with the full complement of

---

[9] Darwin on Trial, p. 133

genes and chromosomes. That means that for all growth within all organisms, every gene must duplicate itself, every chromosome must duplicate itself, and every nucleus of every cell must duplicate itself. The complexity of this process is nothing short of staggering. And it happens constantly as each and every organism grows. The process of cell division is called mitosis, and here's an illustration.

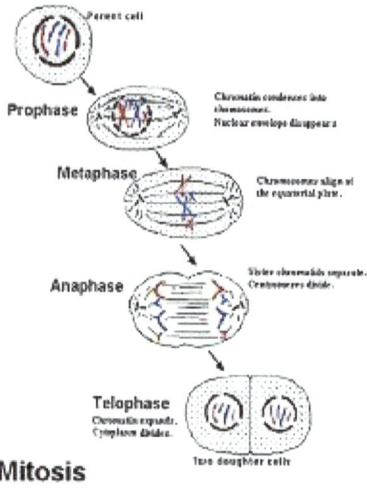

Mitosis

Cycles

Life requires more than just cells, it requires cycles and processes. Food must be digested and absorbed. Oxygen must be breathed in and absorbed, and so forth. One of the most basic cycles required for life is the "Citric Acid Cycle". This cycle involves a series of nine chemical reactions, each reaction leading to the next, one after the other, like a line of dominos, flowing in precise order in a continuous sequence[10]. When my biology professor was teaching on the Citric Acid Cycle, he used a large poster containing all the reactions in a circle, much like the picture on the next page. It took up the whole poster. I posed the question, "How long does it take for one cycle?" He shrugged and said, "About $1/1000^{th}$ of a second."

---

[10] 1. Before the pyruvates from glycolysis can feed into the citric acid cycle, they must undergo a transition reaction. The pyruvate is converted into a 2-carbon acetyl group as the third carbon is lost as CO_2. The acetyl group is attached to coenzyme A to form acetyl-CoA.
2. The 2-carbon acetyl-CoA combines with the 4-carbon oxaloacetate of the citric acid cycle to form 6-carbon citrate.
3. Citrate is converted to isocitrate.
4. The 6-carbon isocitrate is oxidized by NAD^+ to produce reduced NADH and 5-carbon alpha-ketoglutarate. (One carbon is lost as CO_2.)
5. The 5-carbon alpha-ketoglutarate is oxidized by NAD^+ to produce reduced NADH and 4-carbon succinyl-CoA. (One carbon is lost as CO_2.)
6. Oxidation of succinyl-CoA produces succinate and one GTP that is converted to ATP.
7. Oxidation of succinate by FAD produces reduced $FADH_2$ and fumarate.
8. Fumarate is converted into malate.
9. Oxidation of malate by NAD^+ produces reduced NADH and oxaloacetate.

The two molecules of acetyl-CoA from the transition reaction enter the citric acid cycle. This results in the formation of 6 molecules of NADH, two molecules of $FADH_2$, two molecules of ATP, and four molecules of CO_2. The NADH and $FADH_2$ molecules then carry electrons to the electron transport system for further production of ATPs by oxidative phosphorylation.

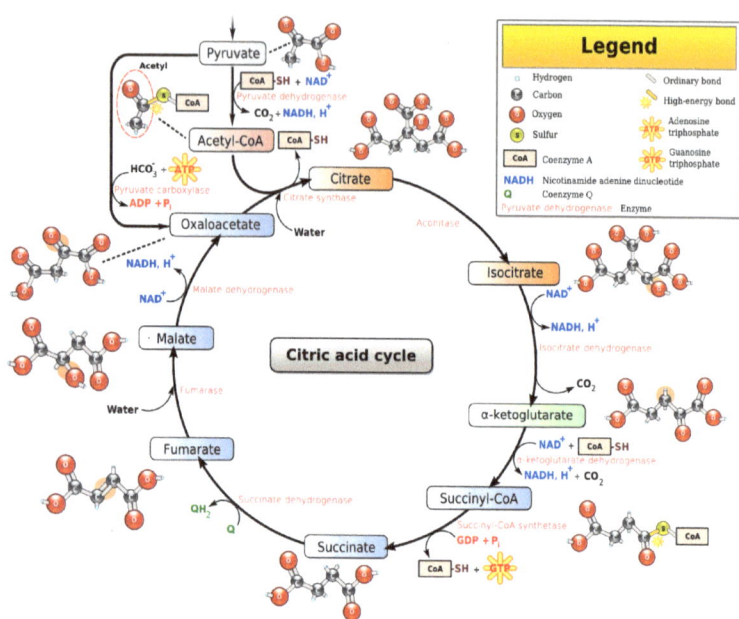

This cycle's complexity is mind-blowing, but in the world of nature, it is just basic physiology and rather routine. And every living thing has dozens of such cycles working continuously.

Plant life has cycles that are no less complex, such as photosynthesis. In this cycle, plants use sunlight to synthesize their food (carbohydrates, sugar) from carbon dioxide and water. It is an indispensable part of a plant's existence, so let's take a look at what is involved.

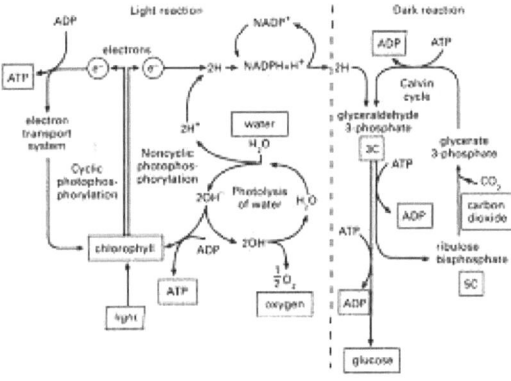

Sunlight, CO2 and H2O are naturally available, but a simple leaf takes them through a series of precise chemical reactions in perfect sequence for the production of food for the plant and oxygen for the whole earth.

Nerves

My Neurophysiology class was taught by the author of our textbook. The entire class was basically devoted to the study of nerve tissue, nerve synapses and nerve impulses. A nerve impulse is pictured as a straight line, a sharp spike up, a sharp drop down which passes below the line and then a straight line again. I could understand the sharp spike up. That made sense. But I could not understand how the nerve impulse could go down below the line. How could it go into negative territory? What was happening? So I posed the question to our professor. Well, you would have thought that I had insulted his mother. The best that I could glean from his quite emotional answer was that it was a very difficult question and that this is a very complicated process

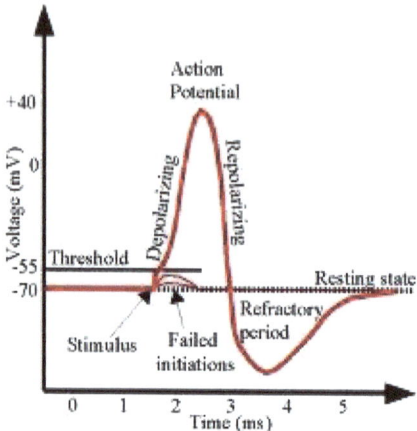

and that there are still a great many mysteries. This was from the author of our neurophysiology textbook.

Perfect Harmony
        Numerous other complicated systems are required to operate in perfect order and in perfect harmony with one another for an organism to function. Each life form's ability to see, breathe, eat, hear, taste, feel, move and so forth requires a complex system to operate. And each system must operate in perfect harmony with all the other systems.

In total
        So the progression of complexity goes like this: subatomic particle to atom, atom to molecule, to gene to chromosome to cell to organ to system to harmony of systems.
        Evolutionists are asking a thinking population to believe that 103 different atoms, thousands of different molecules, genes, chromosomes, cells, organs, cycles, and processes within thousands of

different species came into being in their current perfect operating condition purely by accident.

I don't buy it.

Hooker's Lips

Chapter 4
The Scoop on Scopes

Evolutionists' first line of defense is the assertion that evolution has been accepted by the scientific community for so long that it no longer can be challenged or asked to prove itself. Whenever challenged, evolutionists smugly state that the entire scientific community believes the theory and leave it at that. But the truth is that evolution has never been forced to prove itself and the entire scientific community does not believe it. The famous Scopes "Monkey Trial" does not qualify as a valid test, and here is why.

On March 21, 1925, the State of Tennessee outlawed the teaching of evolution in the public schools, and the newly formed American Civil Liberties Union (ACLU) saw the new law as an opportunity to begin fulfilling its purpose. [11] The ACLU advertised an offer to defend any teacher willing to challenge the new Tennessee law.

George Rappleyea had moved to Dayton, Tennessee from New York and was a staunch evolutionist, but more importantly, he was the head of the Cumberland Coal and Iron Company. This company was the principal employer in Dayton, a town of only 1,800 people, but his company had

---

[11] Roger Baldwin, founder of ACLU, explained its purpose like this: "I am for Socialism, disarmament and ultimately, for the abolishing of the State itself ... I seek the social ownership of property, the abolition of the propertied class and sole control of those who produce wealth. Communism is the goal."

extracted all the coal and iron from the surrounding area, and now his company and the town faced a bleak financial future. When he saw the ACLU's ad, George hatched a plan, and presented his plan to the town's leaders at the local drugstore: Let Dayton be the site for the court battle between evolution and creation and the notoriety would revive the town's struggling economy. The town leaders agreed.

George contacted the ACLU and gave them the good news.

Then, George approached his friend John Scopes while Scopes was playing tennis. Scopes, age 24, was the football coach and math teacher at Rhea County High School, but he could not remember ever teaching evolution. No problem. The town leaders arranged for Scopes to teach biology for one day as a substitute teacher, and he conveniently brought up the subject of evolution. The town constable arrested Scopes and he went back to playing tennis. He never went to jail or endured the slightest bit of scorn for his role in the scheme.

The trial became the trial of the century and a major media event. It was soon dubbed the "Monkey Trial" and the town's ranks swelled to twice its normal size. Radio stations, cameras and the famed journalist H. L. Mencken from the Baltimore Sun, a man known for his irreverence, covered the event. The crowd in the courtroom was so large that county officials feared that the floor in the old courthouse would not hold the extra weight, and the July heat was so intense, that on the second day, the trial was moved outdoors.

Both attorneys who joined this circus were already celebrities. For the defense was Clarence Darrow, a former labor lawyer and the best and most famous trial lawyer in the country, known for his eloquence and atheism. For the prosecution was William Jennings Bryan, "The Great Commoner", three time presidential candidate for the Populist Party (1896, 1900, 1908), former Secretary of State, unofficial national spokesman for Christian fundamentalists, and a man who was also famous for his eloquence.

Much of the trial was heard without the jury present. Judge John Raulston heard all of the defense's scientific "evidence" about evolution without the jury present and ruled that it was irrelevant, citing that the only issue before the jury was whether or not the defendant taught evolution in violation of the criminal statute. Darrow protested so strongly that he was cited for contempt of court.

The most dramatic and most famous scene occurred when the two attorneys squared off, one as witness and one as attorney. Darrow called Bryan, his opposing counsel, to the witness stand to testify as an expert on the Bible. (Testimony by an attorney in a case is prohibited in trial procedure and no attorney can be compelled to do so. One account explains this bizarre move as a deal between the two lawyers: if Bryan agreed to testify, then Darrow would do the same.) Bryan took the witness stand and the two great orators debated each other for two hours. Darrow ridiculed Bryan for believing such Bible stories as Jonah and the whale, the Tower of Babel and Joshua's prayer that the sun stand still. At one point, Darrow yelled at Bryan, "You insult every man of science and learning in the world because he does not believe in your fool religion... We have the purpose of preventing bigots and ignoramuses from controlling the education of the United States." The scientific merits of evolution were scarcely touched. The judge also heard this duel without the jury present, and again ruled that it was irrelevant. Darrow reneged on his end of the bargain and refused to take the witness stand himself.

Most would agree that Darrow came across as the winner of that two hour debate, and for that reason alone, to this day, many believe that evolution beat creation in a court of law.

The journalist H. L. Mencken was ruthless in his reporting on the trial. He portrayed the town of Dayton, and the entire South with it, as a bunch of hicks, and John Scopes as a great martyr. He referred to Bryan as a 'buffoon" and the people of Dayton as 'morons' and 'yokels'.

The trial lasted eight days. The jury was out for only nine minutes and found John Scopes guilty of teaching evolution. Judge Raulston imposed the minimum sentence, a $100 fine, which the Baltimore Sun paid. The verdict was later overturned on appeal on the grounds that the jury, and not the judge, should have imposed the fine.

John Scopes was now quite the celebrity, and Rhea County High School was eager to have him return, but a scholarship had been put together by scientists and journalists to pay for Scopes to study geology at the University of Chicago. So Scope went to graduate school and from there to a successful career in the oil business.

The town of Dayton soon returned to its former sleepy ways and never reaped the great financial rewards that Mr. Rappalyea and the town leaders had hoped for.

Those are the historical facts. Now here is how the Broadway play and Hollywood movie portrayed the event.

"Inherit the Wind" was a 1955 Broadway play and a 1960 movie based on the Scopes trial. The play contained the disclaimer, "Inherit the Wind is not history" and both the play and movie do not use the real characters' names, but everyone who has ever seen the play or movie is led to believe that it is a true story. The movie starred Spencer Tracy as the defense lawyer, Fredric March as the prosecutor, and featured Gene Kelly, Dick York, Harry Morgan, Claude Akins and Noah Beery. The movie was nominated for four Academy Awards.

In the movie, the Scopes character is arrested in front of his class, dragged out of school, thrown into jail where he is harassed by an angry mob led by a vicious preacher, and then publicly burned in effigy. The mob then screams its way over to the hotel where it threatens the teacher's defense attorney.

There is no mention of the ACLU's advertisement, the town's plan, or the teacher's willing involvement. Rather, the teacher is portrayed as the innocent victim of a vicious attack by ignorant and bigoted Christians.

The Clarence Darrow character is portrayed as a man deeply concerned for his persecuted client. In truth, Darrow stated that he took the case for free because he "wanted to show up fundamentalism".

When the Bryan character is testifying as a witness in court, the Darrow character asks him a question about original sin. The Bryan character answers that Adam and Eve's original sin was when they had sexual intercourse. No such exchange took place during the real trial, and William Jennings Bryan would never have said anything so silly.

The Bryan character dies of a dramatic heart attack while giving his closing argument. In truth, Bryan died in his sleep five days after the trial. The real life H. L. Mencken is said to have responded to the news of his death with, "Well, we killed the son of a bitch".

There are numerous other inconsistencies. In general, the Bryan character is portrayed as a hysterical fanatic and the townspeople are portrayed as mean, intolerant, ignorant and intent upon suppressing intelligent thought and scientific truth. While the Darrow and Scopes characters are progressive and intelligent martyrs.

Yet, despite the enormous disparity between historical fact and the movie, and the lack of true scientific debate on the merits of evolution, the movie continues to be viewed as an accurate comparison of the two sides and therefore, any intelligent person must side with Scopes, evolution and intelligent scientific reality.

The lack of integrity so conspicuously displayed in the production of Inherit the Wind is characteristic of the lack of integrity that has permeated evolutionists' behavior during the entire course of the debate.

Red-eyed tree frog

Laughing Bumble Bee Orchid

Chapter 5
The Missing Missing Link

When Darwin's book was first published in 1859, its loudest critics were not the clergy, but rather the geologists and paleontologists who knew the geologic column and fossil record and wondered how evolution could have happened and escaped their notice. But nevertheless, Darwin's book sparked a mad race to find the "missing link" between monkey and man and the Grand Prize awaiting its lucky finder. The zeal that ensued left scientific objectivity in the dust with the fossils.

Many bits and fragments have been put forth as the missing link with great fanfare. Usually, it is nothing more than a single tooth. Every one has been a fake. Unfortunately, the initial fanfare gets all the press, and the exposure of the fraud does not. The much hailed Ramapithicus turned out to be an orangutan. Nebraska man was nothing more than one pig's tooth[12] and Colorado man a horses' tooth. Java man was a gibbon and the famous Lucy just a chimpanzee. Numerous internet websites are devoted to exposing nearly each one of these frauds.

"Piltdown Man" is a good example. In 1915, Charles Dawson (not *Darwin*) claimed that he found

---

[12] Based upon a single tooth unearthed in Nebraska in 1922, the journal Science proclaimed it humanlike and the Illustrated London Times presented an artist's depiction of the humanlike "Nebraska Man" walking. Subsequent excavations of the site unearthed the remainder of the skeleton and revealed that the tooth belonged to a peccary, or wild pig. In 1927, the journal Science printed a retraction admitting it was only a pig's tooth. (Gregory, W.K. (1927). "*Hesperopithecus* apparently not an ape nor a man". *Science* 66 (1720): 579–81.)

part of a skull and jawbone in the Piltdown quarry in Sussex, England, hailed it as the missing link and wallowed in the spotlight for the rest of his life. He called it Eoanthropus dawsoni. It is now universally acknowledged that it was an outright and deliberate fraud. But from 1915 to 1953, it was hailed by some as real evidence of the missing link. There are dozens of books and websites and a PBS documentary that chronicle the Piltdown Man hoax as a hoax.[13]

Every school day, thousands of school children go on field trips to Washington, DC, walk into the Smithsonian Institute's Museum of Natural History and are greeted by a huge mural depicting the transition of monkey to man; the monkey at the left, the man at the right, and all of evolution's dreams in-between.

The magnitude of this deception is staggering. Every impressionable child on a field trip to the

---

[13] One book on the hoax is Ronald Millar's The Piltdown Men. Other authors include Spencer, Weiner, Blinderman, and Walsh. The PBS documentary can be found at *www.pbs.org/wgbh/aso/databank/entries/do53pi.html* PBS "Piltdown Man Hoax Is Exposed"

capital of the United States, led by his revered school teacher, is confronted by this fabricated picture that seals into his mind and psyche the indelible image of his origins. He is taught at this tender age that he descended from a monkey. Most of the world, both children and adult, rightly conclude that it must be a proven scientific fact.

Yet there is no evidence whatsoever of any intermediate form between man and monkey. I submit that there is no other single scientific premise in the entire world for which there is so little hard evidence. Despite Herculean efforts on the part of nearly every paleontologist on the planet, the missing link is still missing. There is not one single piece of credible evidence to support even one of the intermediate forms depicted, and many millions of intermediate steps are required for the theory to be scientific fact. The missing link is still missing, and in their frantic search, evolutionists have established a consistent and proven pattern of dishonesty for which they have never been held accountable. Yet the pictures persist.

Fossils

The Missing Link is not just missing for humans. It is missing for all other life forms as well. Evolution requires that all present life forms, both plant and animal, are the result of trillions upon trillions of minute changes over many billions of years. But if so many changes occurred, then they must of necessity have left lots of evidence behind on the surface of the same planet where all of this happened.

Many people find fossil hunting (Paleontology) fascinating, and as a result, there is now a substantial fossil record, with over 100 million fossils cataloged in museums. However, not even one of these fossils represents an intermediate form that could be put forward as evidence of evolution. That's right, not one. All that's been found are either extinct species or species that are alive today.

Even Charles Darwin admitted this:

"Not one change of species into another is on record ... we cannot prove that a single species has been changed." [14]

Science Researcher Luther Sutherland contacted the heads of five top museums of natural history, including the British, Washington, Field in Chicago and Harvard to determine if any of them had any intermediate fossils in their museums. He found that "None of five museum officials could offer a single example of a transitional series of fossilized organisms that would document the transformation of one basically different type to another."

America's foremost paleontologist, George Gaylord Simpson (1901-1984), admitted this:

"This regular absence of transitional forms is not confined to mammals, but is an almost

_____

[14] The Life and Letters of Charles Darwin, by Francis Darwin,[1898], Basic Books: New York NY, Vol. II., 1959, reprint, p.210.

universal phenomenon, as has long been noted by paleontologists." [15]

Steven Stanley, a paleontologist known for his study of the Bighorn Basin in Wyoming, wrote: "the fossil record does not convincingly document a single transition from one species to another."[16]

Another paleontologist, Niles Eldredge admitted, "We paleontologists have said that the history of life supports (the story of gradual adaptive change), all the while really knowing that it does not."[17]

In his book of this subject, Phillip Johnson wrote, "If Darwinism enjoys the status of an a priori truth, then the problem presented by the fossil record is how Darwinist evolution always happened in such a manner as to escape detection."[18]

The popular science writer Francis Hitching wrote: "But the curious thing is that there is a consistency about the fossil gaps: *the fossils go*

---

[15] Tempo and Mode in Evolution (New York: Columbia University Press, 1944), p. 107.

[16] Darwin on Trial, by Phillip E. Johnson, InterVarsity Press, 1991, page 51

[17] Ibid, page 59

[18] Ibid, page 53

missing in all the important places." (emphasis in original)[19]

Newsweek Magazine reported, "The more scientists have searched for the transitional forms that lie between species, the more they have been frustrated." [20]

Even evolution's darling, Dr. Stephen J. Gould of Harvard, America's most prolific writer in support of evolution, said "the failure to find a clear vector of progress in life's history as the most puzzling fact of the fossil record."

He also said "the extreme rarity of transitional forms in the fossil record is the trade secret of paleontology."[21]

It is such an obvious problem for evolutionists that Dr. Gould submitted a paper at the world meeting of evolutionists at the Chicago Museum of Natural History stating that "it has long been a trade secret among paleontologists that there are no intermediate forms and the time has come to admit this publicly". [22]

But as spokesman for the cause, Dr. Gould felt the need to explain away the problem and so came up with what has become his most famous statement. It is found in his 1977 book, Ever Since Darwin:

---

[19] *The Neck of the Giraffe: Darwin, Evolution and the New Biology,* 1982, pp. 9-10, by Francis Hitching, member of the Prehistoric Society and the Society for Physical Research.
[20] John Adler with John Carey, "Is Man a Subtle Accident?", Newsweek, Vol.96, No.18 November 3, 1980, p.95
[21]Ibid, page 59

[22] October 1980

"Science is not a heartless pursuit of objective information. It is a creative human activity, its geniuses acting more as artists than as information processors."

Evolutionists derive enormous comfort from Dr. Gould's line of reasoning. As spontaneously as evolution itself, science no longer required objective information, but could put creativity and artistry in its heartless place. For therein lies true genius.

But what actually lies therein is evolution's most revered spokesman admitting the vast vacuum that is evolution. Otherwise, there would be no need to substitute creativity and artistic genius in the place and stead of actual facts.

Paleontologists are out there digging as you read this, but they don't bother to tell anyone about all the fossils of extinct species or existing species that they continue to find. Because all that matters to them is finding the missing link between anything and anything else, or any bit that can be hailed as such. They haven't found it yet.

Geologic Layers

One of the biggest lies perpetrated by evolutionists is the fabricated picture of the different layers of the earth's crust showing fossils of simple forms at the bottom and increasingly more complex forms as you progress up. Perhaps you have seen these pictures. One was in my biology textbook. They are a total fabrication.

Dr. David Raup, Curator of Geology at the Field Museum of Natural History in Chicago said "Most people assume that fossils provide a very important part of the general argument in favour of Darwinian interpretations of the history of life. Unfortunately, this is not strictly true."

Not all of the earth's crust contains fossils. There is only a thin layer of crust that contains any fossils at all and it is sporadically scattered around the earth. There are no sites where there are simple fossils at the bottom, complex at the top and intermediate in-between. The pictures are bogus.

> "Eighty to eighty-five percent of earth's land surface does not have even 3 geological periods appearing in 'correct' consecutive order ... it becomes an overall exercise of gargantuan special pleading and imagination for the evolutionary/uniformitarian paradigm to maintain that there ever were geologic periods." [23]

Summation

Fossils exist. There are lots of them and at least some of them should document the billions of transitional forms required by this theory. But they don't. Why is that? If anyone wants hard scientific proof that evolution never occurred, then this is it. A single pig's tooth won't cut it.

---

[23] John Woodmorappe, - Geologist

Chapter 6
Probability

Mathematicians can calculate the probability of things. I took a Probability Course at UVA and all current college curricula for statistics include a course on probability. It is a fully developed and recognized field of mathematics and well equipped with formulas to calculate the probability of something. Any poker player can tell you that. In mathematics, the laws of probability are as reliable as 2 + 2 = 4.

Evolution is based on the premise that all living things came into being totally by accident. That includes atoms, molecules, genes, chromosomes, cells, tissues, organs, systems and life itself, within every living thing. To discuss the probability of this, we will skip over the marvel of the atom for now and look at the next simplest ingredient, the molecule.

The most basic ingredient in both plant and animal biology is the protein molecule which consists of over 300 atoms in specific order and alignment. It is the building block of all living things.

Of course, mathematicians could not resist calculating the probability that a protein molecule came into being by accident and many have done so. As a starting place, 200 working parts was chosen, because even the simplest living molecule in nature has *more than* 200 atoms in proper configuration. So here is the math question: What is the probability that 200 atoms fell together by accident in just the right order to make up one protein molecule? The answer: the probability is 1 in a number with 375 zeros. (A trillion, for instance, has 12 zeros.) It would require a billion tries per second for over 90 billion years for it to happen only one time. In mathematical terms, an absolute zero. And for that calculation, only 200 was used, whereas the most basic protein molecule has over 300 atoms in precise configuration.

One way to visualize this is to write out the first sentence of the Declaration of Independence, 330 letters, using an individual block for each letter.[24] Lay all the blocks out on the table just right so that it spells out the sentence. Then scoop up all 330 blocks and throw them down on the table. What is the probability that they will line up just right so as to say the sentence just like Thomas Jefferson wrote it? That is what must happen for a 330 part molecule to come together by accident.

Marcel Golay, a noted mathematician, calculated the probability of the simplest protein molecule coming together by accident. Its probability is one in 10 to the 450th power, or one in a number with 450 zeros.

DNA is a necessary and indispensable molecule in nature. So a more viable test for evolution is the probability of a DNA molecule forming, rather than a hypothetical part with only 200 or 300 ingredients. So the noted mathematician, R.L. Wysong, calculated the probability of a DNA

---

[24] "The unanimous Declaration of the thirteen united States of America, When in the Course of human events, it becomes necessary for one people to dissolve the political bands which have connected them with another, and to assume among the powers of the earth, the separate and equal station to which the Laws of Nature and of Nature's God entitle them, a decent respect to the opinions of mankind requires that they should declare the causes which impel them to the separation."

molecule coming together by accident. It is one in a number with 167,626 zeros. To give the reader a

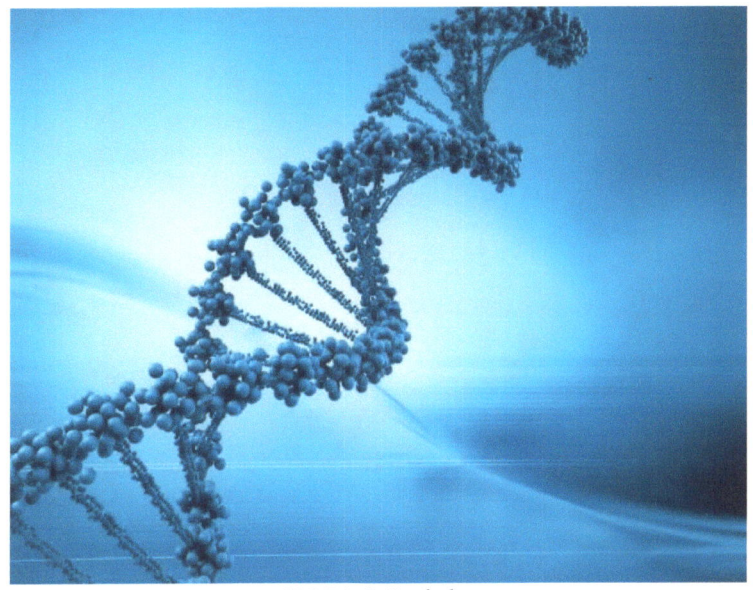

DNA Model

frame of reference for just how large this number is, consider that there are 10 to the 70th power (a number with 70 zeros) atoms in the entire observable universe.[25] Ten to the 167,626th power is 2,400 times larger than that. In mathematical terms, it can't happen, ever, not even once.

Dr. Francis Crick, the co-discoverer of DNA, computed the probability of one DNA molecule

---

[25] Estimates also include:
300-500 billion galaxies and 1.2 x 10 to the 23rd power stars in the observable universe. Using these figures, the total number of atoms in the universe is estimated at 10 to the 80th power. "How many atoms are there in the universe?" by John Carl Villanueva, 2009.

coming into existence by chance over 4½ billion years. Its probability is zero.

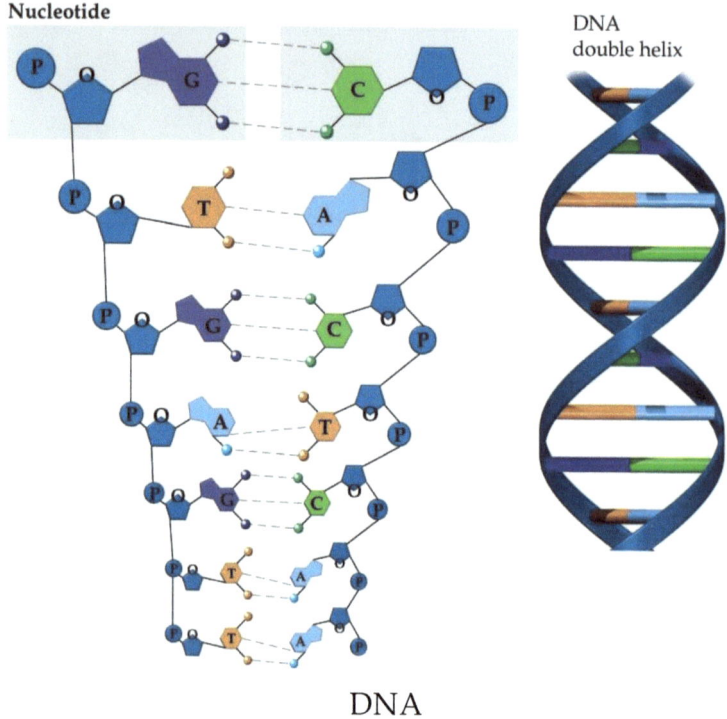

Nucleotide

DNA double helix

DNA

Sir Fred Hoyle calculated the mathematical probability of a single cell coming together by chance anywhere in the universe in 20 billion years. The probability is zero.[26]

---

[26] "The likelihood of the formation of life from inanimate matter is one to a number with 40,000 noughts after it." Sir Fred Hoyle, - Astronomer, Cosmologist and Mathematician, Cambridge University

Now consider that "the set of genetic instruct-tions for humans is roughly three billion letters long."27

Structure of a Generalized Cell

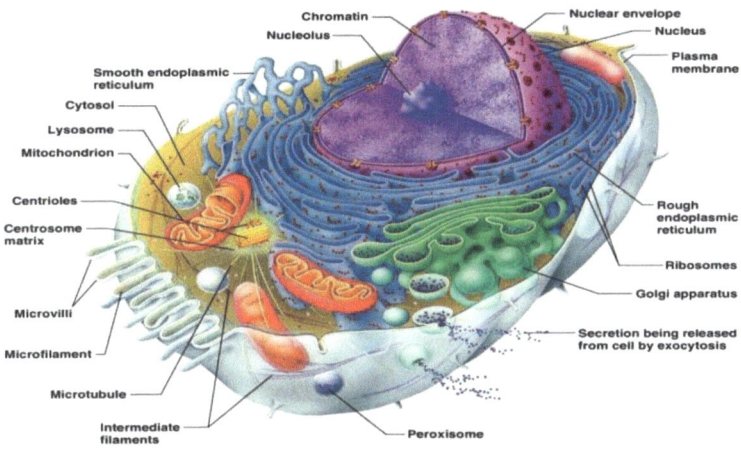

The probability that the simplest single-celled life form came into being through evolution can be compared to the probability of a tornado going through a field containing thousands of separate airplane parts and leaving behind a fully operational 747 airplane, with the engine running. Every calculation of the probability that even the simplest living organism came into being by accident has

---

27 Miroslav Radman & Robert Wagner, - "The High Fidelity of DNA Duplication", Scientific America, Vol. 259, No.2 August 1988, pp40-46

always been the same; an absolute zero. It is a mathematical impossibility.

After considering the probability that cells occurred by accident, it is necessary that we consider how infinitely more improbable it is that such complex functions as sight, hearing, feeling, taste and touch could do it, and how such complex systems as the digestive, cardiovascular, muscular and nervous systems could possibly appear by accident. Evolution requires that we believe that all such things came into being spontaneously by accident.

Just two years after his book came out, Charles Darwin wrote a biologist friend, "The eye to this day gives me a cold shudder." [28]

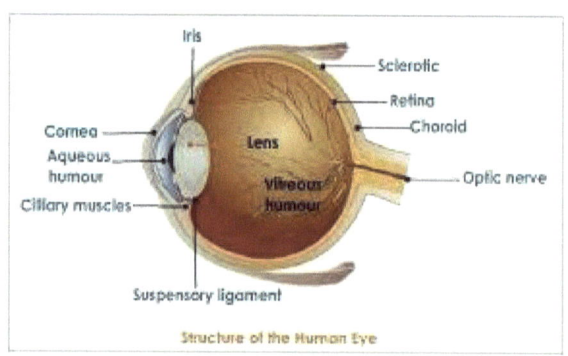

Structure of the Human Eye

Darwin also admitted that "If it could be demonstrated that any complex organ existed which could not possibly have been formed by numerous,

---

[28] Darwin, Charles (1860) in letter to Asa Gray in Life and Letters of Charles Darwin (1888) 3 vols, ed F. Darwin, John Murray, London, vol 2, page 273

successive, slight modifications, my theory would absolutely break down."[29]

---

[29] The Origin of Species, Chapter 6, Charles Darwin.

Chapter 7
The Impasse

Evolution must of necessity occur at the gene level. Genes are the tiny particles that contain the genetic code of every living organism. Despite their incredibly small size, [30] it has been estimated that there is enough information in one gene to fill 750 large books.

Evolution is based entirely upon the premise that genes routinely mutate, or accidentally change, and the mutation is passed on to viable offspring.

Gene mutations are most commonly caused by two types of occurrences; either environmental factors such as chemicals or radiation or ultraviolet light from the sun. These factors can alter the DNA sequence and can even change the shape of DNA.

Other mutations are caused by errors made during cell division for growth and maintenance (mitosis) or during cell division for sex cell production (meiosis).

Sex cells are produced in the body in a process called meiosis. In this process, the chromosomes in the sex cells divide into two parts to form two sperm cells or eggs with half the normal number of genes and chromosomes. (plants have different sex cells, usually spores) The meiosis process is very complex and occasionally it goes awry and the resulting sex cell will have a mutated gene. It is rare, but it does happen.

---

[30] the size of genes is measured in picograms, which are trillionths of a gram

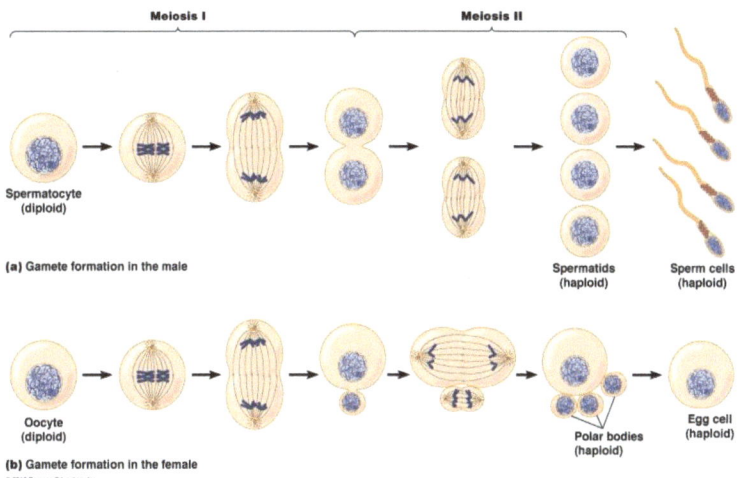

Meiosis I   Meiosis II

Spermatocyte
(diploid)

(a) Gamete formation in the male

Spermatids
(haploid)

Sperm cells
(haploid)

Oocyte
(diploid)

(b) Gamete formation in the female

© 2012 Pearson Education, Inc.

Polar bodies
(haploid)

Egg cell
(haploid)

Meiosis

An embryo forms when a sperm and egg meet and all the male genes on every chromosome successfully pair with every female gene. This is just as true for mutant genes as any other gene. But because a mutant gene has a different shape, it usually has trouble pairing with its counterpart gene, because it is different.

Imagine the male half-chromosome lining up with the female half-chromosome to form the complete chromosome of the offspring. The two halves come together to make one chromosome in the embryo. Each gene lines up with its corresponding gene on the other half and they come together by means of a quite intricate and complex coupling. But alas, it doesn't line up. One gene is different than the other gene on the opposite half. But in order for the two halves to connect, every gene must connect with its counterpart. But it is different and almost always does not connect and therefore no new embryo is produced.   The process fails. This is why most mutations are fatal.

Many experiments have been conducted in an attempt to produce a beneficial mutation. Most of these experiments have been with the fruit fly, because it has huge chromosomes in its salivary glands and can be more easily observed. (Biology students get pretty tired of hearing about the *Drosophila melanogaster.*)

Geneticists began breeding and studying the fruit fly around 1900 and they recorded their first mutation in 1910. Over the next 60 years, they recorded about 3,000 mutations. However, all of the mutations were either fatal, harmful or harmless and none of them produced a more successful fruit fly. [31]

> "In all the thousands of fly-breeding experiments carried out all over the world for more than fifty years, a distinct new species has never been seen to emerge ... or even a new enzyme." [32]

Overall, 98% of all mutations in nature are either fatal, make no significant difference, or are lost in the sea of loneliness. If the mutant does survive, it faces the problem of finding a suitable mate. Just because it survived, does not mean it will find another creature that will accept his chromosome's

---

[31] Dan L. Lindsley and E.H. Grell, *Genetic Variations of* Drosophila melanogaster, Carnegie Institution of Washington, Publication No. 627, 1967

[32] The Great Evolution Mystery by Gordon Taylor, New York: Harper and Row, 1983, pp 34, 38

configuration. The difference is guaranteed and the impasse assured. It cannot produce offspring.

This is why every honest biologist admits that no one has ever been able to produce a single instance of a beneficial mutant finding a mate who passed on his mutation to their offspring. Even the most diehard Darwinists must admit that they cannot produce a single example, and it is not for lack of trying.

> "It is good to keep in mind...that nobody has ever succeeded in producing even one new species by the accumulation of micro-mutations."[33]

It is impossible to exaggerate the significance of this fact and the insurmountable problem this presents for the theory of evolution.

---

[33] Prof. R. Goldschmidt, - PhD,, DSc Prof. Zoology, University of Calif. In "Material Basis of Evolution", Yale Univ. Press

Chapter 8
In a Class by Itself

Normal Standards of Proof

In the scientific community, each of the several scientific fields is filled with people who are wholly devoted to their field of study, even love it, and who strive to discover new things, make some breakthrough that will revolutionize life, and hopefully fulfill their personal destiny and advance their career at the same time. So, there is no shortage of incentives, or of participants, who eagerly labor over their research with the hope of presenting their findings with great fanfare to their particular scientific community. But before any of these hopeful Nobel Prize laureates can grab the brass ring, his research and conclusions must pass the intense scrutiny of his peers; "peer-review". That is the normal standard of proof and it is not easy. Most new "discoveries" are met more often with skepticism than enthusiasm, if not open hostility. Competition and jealousy have proven to be stronger forces of human nature than graciousness. Thus the pressure is on to prove, prove and prove some more.

Louis Pasteur is a rather typical example. His genius was the scorn of the French medical community and every step of progress was in spite of his colleagues, rather than because of them.

But all such normal behavior and standards of proof have played no part in evolution's saga. All normal standards were and have been summarily ignored and Charles Darwin was catapulted to instant celebrity status without even the most casual scrutiny of his revolutionary theory.

Although Charles Darwin wrote <u>The Origin of Species</u> in 1859 when the field of microbiology was nonexistent, and his description of a single-cell organism as "simple" was not as outlandish as it is today, his theory did not then, nor has it since, ever encountered the normal scrutiny that all other "discoveries" have.

The Evolution Express took off at such a high rate of speed and has continued roaring through human thought with so much momentum that it has not needed to account for itself at any time, in spite of the fact that enormous advances have been made in all relevant fields of science and all this new knowledge has challenged the theory at every level possible.

> "The fact that a theory so vague, so insufficiently verifiable, and so far from the criteria otherwise applied in 'hard' science has become a dogma can only be explained on sociological grounds."[34]

Although the evolutionists would have the world believe that evolution is unanimously accepted by the entire scientific community, that is far from the truth. There are thousands of unbiased, brilliant and respected scientists who have applied the normal standards of proof to the theory and judged that it

---

[34] Ludwig von Bertalanffy, - Biologist

fails every exam. I have included an Addendum at the end of this book setting forth a large number of quotes by renowned scientists who will verify my position in a very convincing manner. They have taken every single assertion and premise behind the theory and analyzed the "evidence" in support of it and found that there is no scientific proof whatsoever. NONE! In fact, the evidence actually conflicts with the theory, and thereby challenges the integrity of those who endorse it. Here are some of the opinions of those highly revered scientists:

"The theory of Evolution suffers from grave defects, which are more and more apparent as time advances. It can no longer square with practical scientific knowledge." [35]

"We have had enough of the Darwinian fallacy. It is time that we cry: 'The emperor has no clothes.'" [36]

"Darwin's theory of natural selection has never had any proof, yet it has been universally accepted."[37]

"The theory of the transmutation of species is a scientific mistake, untrue in its facts, unscientific in its method, and mischievous in its tendency."[38]

---

[35] Dr. A Fleishmann, - Zoologist, Erlangen University

[36] K. Hsu, - Geologist At The Geological Institute At Zurich

[37] Prof. R. Goldschmidt, - PhD, DSc Prof. Zoology, University of Calif. "In Material Basis of Evolution", Yale Univ. Press

"Evolution is baseless and quite incredible." [39]

"The Evolutionist thesis has become more stringently unthinkable than ever before." [40]

"A growing number of respectable scientists are defecting from the Evolutionist camp ... moreover, for the most part these 'experts' have abandoned Darwinism, not on the basis of religious faith or biblical persuasions, but on scientific grounds, and in some instances, regretfully." [41]

So, I can assure you that I am not alone in my criticism and the vast array of scholars who have written on the subject have done a much more technical and detailed job of explaining the scientific problems with evolution than I do here. But most of these books use highly technical language, mountains of data and take the reader through complex scientific analysis. This book is intended to make the same point, but by simpler means.

---

[38] "Methods of Study in Natural History", by Harvard Professor J. Agassiz

[39] Dr. Ambrose Fleming, - President, British Assoc.
Advancement of Science, In "The Unleashing of Evolutionary Thought"

[40] Wolfgang Smith, - Ph.D., Physicist and Mathematician

[41] ibid

All they Got

Evolutionists claim that an extinct flightless bird had a skeleton similar to a dinosaur so it qualifies as an intermediate form between reptiles and birds. But it was a bird, very similar to other flightless birds, and not a reptile. They also point with enthusiasm to an extinct small horse as an intermediate form for the modern horse. But a small horse is still a horse and not an intermediate form between two species.

Enormous fanfare goes to the tailbone in humans as evidence of a former tail. However, the tailbone, or coccyx, is quite functional.[42]

They also get very excited about folds that appear to be gills at one stage of the development of the human embryo. However, the folds that have been compared to gills are not located on the area of the embryo that is related to breathing at all.[43] Nor are these folds present during the vast majority of the embryo's growth.

Such are the extreme lengths to which they must resort. And that's all they got.

"The theory of Evolution ... will be one of the great jokes in the history books of the future. Posterity will marvel that so flimsy and

---

[42] "The coccyx serves as an attachment site for tendons, ligaments, and muscles. It also functions as an insertion point of some of the muscles of the pelvic floor. The coccyx also functions to support and stabilize a person while he or she is in a sitting position." *Healthline*

[43] Rather than having anything to do with vestigial apparatus for breathing or respiration, the pharyngeal "pouches" develop into parts of the face, neck, and important glands.

dubious an hypothesis could be accepted with the incredible credulity it has." [44]

Experiments

In their pursuit of proof, evolutionists have conducted many experiments; but to no avail. All experiments with animals and plants to force a transition between species have failed. It is simply impossible for animals or plants of different species to mate with one another and produce fertile offspring.[45] And evolution is not possible without this most fundamental of functions. Therefore, if one hypothetical animal were to evolve into a higher form, it would have no partner to mate with and die a lonely spinster.

> "My attempts to demonstrate Evolution by an experiment carried on for more than 40 years have completely failed." *N. H. Nilson, -Famous Botanist and Evolutionist*

Motive

The bottom line is that there is no living, fossilized or experimental evidence for evolution. But in the guise of science, evolution has presented a wonderful opportunity for the atheist. In their own words, "It finally became intellectually acceptable to be an atheist." Creation has been such an anathema to the godless that they will jump at any opportunity

---

[44] Malcolm Muggeridge, -Famous Philosopher

[45] A horse and donkey can produce offspring, a mule, but it is sterile.

to remove this bone from their throats. A door opened that has attracted a new breed of ideological fanatics and the ranks soon filled with those more committed to their personal freedom than to science.

Whereas evolutionists are restricted to repeating the same tired nonsense over and over, real scientists have an ever growing mountain of evidence that contradicts the theory.

At our universities, the most important measure of success and status among the faculty is publishing. Perhaps you have heard the old saying, "Publish or Perish". It is true. It does not refer to textbooks or books sold at Barnes and Noble. It refers to articles published in academic journals. So most university professors are doing everything possible to get published in their field's major journals, and the competition is fierce. If any professor is known to be a creationist or anti-evolution, then his chances of being published are slim or none. This is a powerful force in academic circles. It is more peer pressure than a teenager choosing his wardrobe for the first day of school. All professors and their doctoral students must choose between creation and their career. It is that simple.

Piliated Woodpecker

Chapter 9
The Second Law of Thermodynamics

In science, there are some truths that are so fundamental and indisputable that they are given the designation of "laws". One of them is the "Second Law of Thermodynamics", the law of energy decay. I studied it in physics. It states that in every system left to its own devices, heat will only move from a hotter object to a colder object, and not the other way around, and that order will only move to disorder, and not the other way around.

For example, if you placed a silver dollar with a temperature of 100° on top of a silver dollar with a temperature of 50°, what will happen? Will the heat in the colder silver dollar move to the hotter silver dollar and make it 150°? No, of course not. Heat will move from the hotter to the colder, and you will eventually have two silver dollars at 75°. This is such basic science that it has earned the esteemed status of a scientific "Law".

This same Second Law of Thermodynamics holds that all things in the universe left to themselves will degenerate from order to disorder. It is a basic principle of nature and universally accepted by the entire scientific community as a fundamental "law of nature. "

The theory of evolution relies upon the premise that living organisms spontaneously erupted out of inert matter and then progressed from simple to complex, over and over, trillions upon trillions of times. Therefore the theory of evolution is in direct conflict with this basic law of science. In scientific terms, it cannot happen. To many an objective scientist, this alone has been enough to discard the entire theory, and they have.[46] They cannot accept the premise that the natural order of the universe forced a lifeless particle to become a man.

[46] Lambert, F (2002). "Disorder — A Cracked Crutch For Supporting Entropy Discussions". *Journal of Chemical Education* **79** (2): 187–192. Bibcode:2002JChEd..79..187L. doi:10.1021/ed079p187. Retrieved 2007-03-24.

Chapter 10
A Wall

As we have covered, the heart and soul of the evolution theory is the passing on of beneficial changes, one gene mutation at a time. Then after several millions of these small changes, a new species emerges. This process requires that the new life form be able to survive and reproduce after each gene mutation. There is no place in evolution for any step along the way that cannot perpetuate the change. For instance, if there was any step in the process that required several million genes to change all at once in order for that form to reproduce, then that form is not possible in the evolutionary scheme. Such a requirement would erect a wall so high that no amount of evolutionary imagination could possibly leap over it.

But in nature, there are several species that exhibit intermediate forms that have erected such an insurmountable wall. Frogs, butterflies and mosquitoes present good examples.

With the butterfly, there is an intermediate caterpillar form which hatches from an egg, eats leaves, weaves a cocoon, and while in the cocoon, turns into a butterfly. The problem this presents for evolution is that the caterpillar stage cannot reproduce. Therefore if this species evolved, the entire caterpillar/cocoon stage had to emerge instantaneously. Conservatively speaking, this would require several million mutations at one time in one butterfly

for evolution to leap over this wall; which is beyond the scope of even the wildest scientific imagination.

The exact same problem exists for the evolution of the frog, which goes through an intermediate tadpole stage. Tadpoles cannot reproduce.

Mosquitoes

It is also true with mosquitoes. I hate mosquitoes as much as anyone, but they play a crucial role in nature's balance and provide another example of the problems with evolution.

As we will soon read, the beaver pond is the hub of forest life. But when the pond first forms, the only wildlife in it is that which lived in the stream. It has no large fish.

The large fish arrive by means of their eggs stuck to the legs of visiting aquatic birds. But when these eggs grow up into large fish, they will need something to eat. That is where mosquitoes come in.

Mosquitoes lay eggs in all stagnant water which hatch into mosquito larvae and pupae, millions of them, each one a small wiggly bit of pure protein, dancing about in the middle of the water, just asking to be eaten by hungry baby fish.

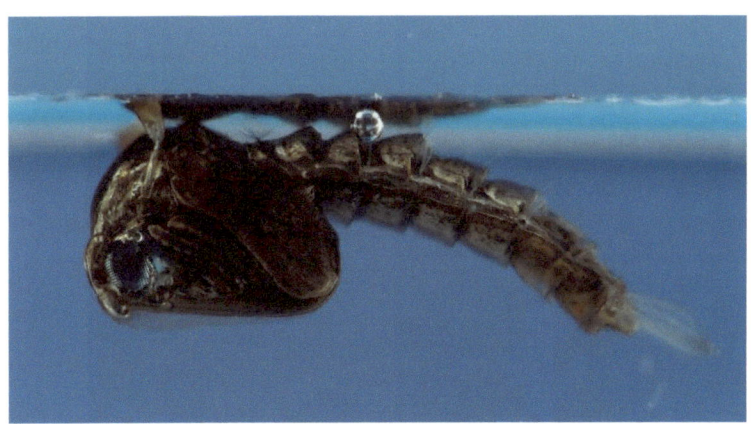

Mosquito pupae

Mosquito larvae

So as unpleasant as it is for animals in the forest, protein is transferred from the land, where there is an abundance of protein, to the pond where it is much needed, and transferred in a very convenient form indeed. This transfer of protein continues to bless every pond during the pond's entire life.

This pesky mosquito is a marvel of blood sucking engineering and a real problem for evolution. It has all the right equipment, including a long straw-like snout that would be of no use whatsoever until fully formed.

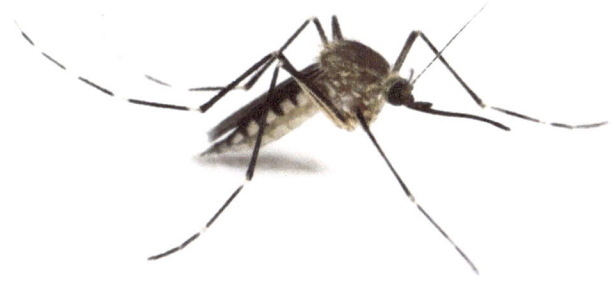

Every mosquito goes through three different stages: larvae, pupae and adult. Each stage is necessary to produce an adult, and only an adult can lay eggs for reproduction. So how did the first two Stages survive over millions of years until they got to the adult form capable of reproduction? The larval and pupae stages cannot reproduce, so the mosquito's evolution must have, of necessity, stopped with the larval stage.

In addition, the first two stages live under water and the adult lives above water. The water dwelling stage must change into an air breathing stage in a matter of seconds or die. The papal stage does not have millions of years to wait for just the right accident to occur on its chromosomes. It only has a few seconds. The evolution of this species is scientifically impossible.

Beavers

When I was very young, I read a book about a family that moved to a remote area of British Columbia, Canada in the 1930's to experience the wilds of nature. [47] But after they got there, they discovered to their surprise that although it was wilderness, there *was* no nature. There were very few animals in the entire forest. So they asked an elderly Indian woman about it. Her answer was simple, "No beavers. The Indians trapped them all to trade for drink and trinkets". She said that most animals do not live in a pure forest, no matter how remote, but rather on the edge of the forest where it meets a meadow, in meadows, and in and around ponds. There were no animals because there were no meadows and no ponds, and there were no meadows or ponds because there were no beavers. She explained that beavers build the dams that create ponds. These ponds then provide a habitat for abundant pond life including fish, frogs, crawfish, muskrats, ducks and cranes. But they also flood the area and the flooding kills all the trees in the pond. Then after some years, the pond fills up with silt, the beavers move on and the dam breaks down. After the pond drains, a very fertile field is left behind and, voila, you have a meadow. She said that all the beavers had been trapped in that area, and without beavers, there was no wildlife, only woods.

---

[47] Three Against the Wilderness, by Eric Collier. Published in 1959 to. Eric died in 1966; their son Veasy died in 2012.

So the family decided to bring beavers back to their woods. The book was their very interesting story and become one of the most famous books ever written about British Columbia.

The beaver has all the right equipment and instincts for what it does. It loves to cut down trees and build dams. It provides an invaluable service to all the other forest life, but what evolutionary forces drove this rodent to become so very helpful? Did all the wolves, coyotes, foxes, bobcats, wolverines and cougars get together and agree that they would not eat this rodent for a few million years so that it could evolve into a dam building engineer for the good of the forest?   Or did this rodent jump from rat to engineer in a single bound? All intermediate forms would not be the engineering and wood cutting marvel that a beaver is today. Why would it grow that strange tail? It would not have a pond for protection and would not have survived one day, not to mention millions of years.

Hippopotamus

This animal is big and fat and lives in Africa's rivers. When on land, it is fast, but vulnerable against larger predators. So it spends most of its time in the water where it is safe from lions and leopards and its size is not a liability. So, it waddles around in Africa's rivers with the crocodiles. Crocodiles!!! Yes, 20 foot crocodiles, that wait for anything to make the mistake of coming into their river. But it leaves the hippos alone. You see, the hippo, though a vegetarian, has huge pointed teeth, about one foot long, perfect for piercing the underbelly of crocodiles and ripping their guts out. The hippos and the crocs swim side by side in all the rivers of Africa without any problem. But how did this happen?

According to evolutionists, reptiles came first, so the crocodiles came before the hippos. So, according to Darwinists, the rivers of Africa had crocodiles in them millions of years before any hippo

came on the scene. Then after some reptile became a bird and some bird become a mammal, hippos came along. And according to evolution's scenario, hippos evolved from other land mammals and moved their habitat from land to river. So the obvious question is how and why did a vegetarian land animal develop huge sword-like teeth that are only useful if it is living in the rivers? Or if it is living in the rivers, how did it survive for the millions of years that it would take to develop those teeth? The scenario is totally implausible; ridiculous in fact.

Whales

Evolution proposes that life progressed in the following sequence:

ooze to fish

fish to amphibians

amphibians to reptiles
reptiles to birds
birds to mammals

Now consider that whales are mammals. They give birth to live young and nurse them with milk. But according to evolution, there were amphibians, reptiles and birds before mammals. So, according to evolution's own timetable, *a land mammal evolved into a whale, the largest of all living things.* Are you kidding me?

Blue whale; 80' long, 150 tons

In his book <u>Darwin on Trial</u>, Philip Johnson repeatedly challenges the reader to explain how a whale and a bat, both mammals, could have evolved from the same original mammal.

All thinking people must ask the same question. Evolution requires that they did. To me, it is impossible to exaggerate the level of foolishness necessary to believe such nonsense.

Bears Hibernating

Bears are as hungry as... a bear. Their appetites are legendary. They need to eat constantly and if they had to eat during the winter, it would be a big problem, because there is not enough food to sustain them during the winter. So they sleep, all winter.

It is normal for most animals to sleep at night. They can make it that long without eating, drinking and eliminating waste. But after a normal night's sleep, they are more than ready to eat, drink and do their business. So how does a bear sleep all winter without eating, drinking or eliminating waste?

When a bear hibernates, its body undergoes a remarkable change. It is not just sleeping; it enters a very deep sleep where its metabolism slows to a crawl and its urine is recycled. So he is able to sleep for many months without a problem.

But how did this unique process happen? The process of recycling urine is so extraordinary and complicated that it challenges human understanding. One must be severely deceived to think it could happen by accident at all, but even if one considered a bear adapting to this need, one must question how it could happen in thousands of minute steps. The final product is needed for the bear to live through the winter, not 10% or 20% or even 90%. The bear doesn't have that long to wait. Its ready for bed.

Other Examples

Another example is the bombardier beetle. This little bug squirts out an irritating, odious and explosive gas at its attackers. The chemicals used for this concoction are stored separately in the beetle in special storage tanks which keep the dangerous chemicals apart. When danger comes, the beetle mixes the chemicals; they explode and squirt out at the attacker. Now how could this process develop in a step by step fashion? Even the slightest deviation from the precise procedure and our little beetle is a goner, blown to bits. It could not have developed in steps.

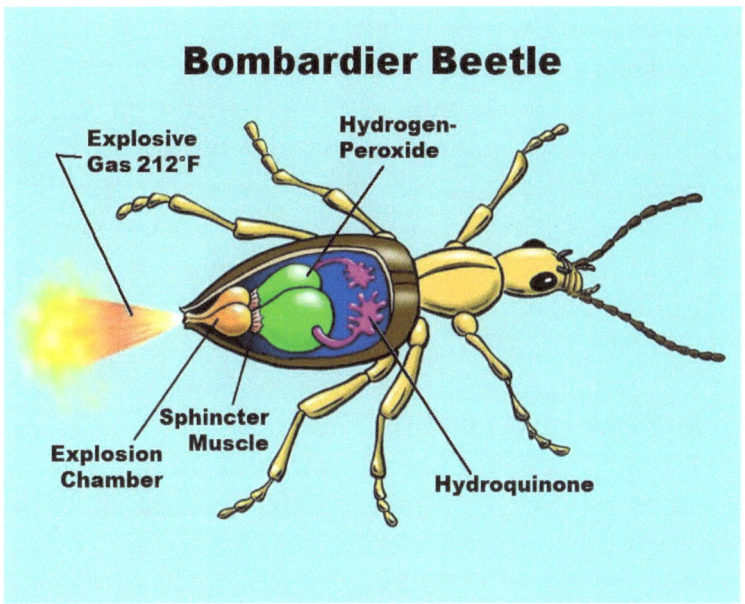

The same is true of the woodpecker. Its head can withstand an impact that would kill any other bird, but for it, it needs this special head to eat. [48] It could not develop in stages and still find those buried bugs.

The bird wing is another example. It could not develop in gradual steps and still be useful. Every

---

[48] The average woodpecker pecks 12,000 times a day, and when they really get going they can reach speeds of up to 20 pecks per second — that's 1,200 pecks per minute. But with a special hyoid bone around the brain, 99.7% of the force energy the woodpecker feels the instant it makes contact with a tree is absorbed by its body.

single step along the transitional road to actual flight would be a hindrance, not an advantage.

The eye presents the same problem. For the eye to see, the eye must work with nerves and the brain. And in many instances, the simpler life forms have more complex eyes than humans. But in all cases, the eye only sees if it works in tandem with the nerves and brain. There is no motivation for only part to exist. They must all appear simultaneously for the eye to function at all.

No room for survival of the fittest

There are countless examples in nature of creatures whose very existence is inconsistent with "survival of the fittest". Look at this *slug* and *worm* and tell me what evolutionary forces drove them to develop such intricate beauty.

Blue Angel Slug

Christmas Tree Worm

How realistic is it to believe that the mating rituals of these slugs and worms were so demanding that it forced them to become the exquisite creatures that they are?

Mutual Symbiosis

Mutual symbiosis is when two unrelated species coexist for the mutual benefit of both. Like ants and aphids, bees and flowers, and algae and fungus.[49]

Coral reefs are formed by coral which is an animal with a limestone skeleton. It obtains 90% of its food from algae, a plant, which lives inside the coral. Coral could not live without algae. So the two, one a plant and the other an animal, needed to evolve step by step and side by side in order for coral to come into being. What are the chances of that?

---

[49]Alga and fungus live together to form lichen which differs from both. Fungus has no way to produce its own food, but gets it from the alga. Alga cannot get its own water, but gets it from the fungus.

Simple Observation

Even the most casual examination of nature reveals that every species is in a state of being that is perfectly suited for where they live and what they do, their "niche" [50]. They have their own place in nature and no motivation to change. In fact, any change at all would doom their existence.

Rather than describing each as having "adapted" to its environment, couldn't we more accurately described each species as being "perfectly suited" to its niche.

Observation of nature also discloses that no species is in transition, but instead, each has reached the pinnacle of form necessary to succeed. None is dragging a leg. Isn't it curious that if evolution is an ongoing process, that it is no longer ongoing?

For the entire 6,000 years of recorded history, each of the millions of living animals and plants on earth has reproduced every year. How many total offspring is that? And not one species has changed.

Just think about the first animals that pop into your head. Are they not perfect for where they live and what they do? They have no reason to change to succeed. Have you ever watched a squirrel scurry up and down a tree, or a fish swim? Do they appear to be struggling to become something different?

-----

[50] niche: "the function of an organism within a community". The World Book Dictionary, World Book, Inc., 1992
"a habitat supplying the factors necessary for the existence of an organism or species." Merriam-Webster Dictionary (biology) (ecology) the ecological role of an organism in a community especially in regard to food consumption.

large mouth bass

Chapter 11
A Lousy Date

Evolution requires that the earth be many billions of years old, and therefore evolutionists advance that belief. As a result, many a normal person feels that his intelligence is suspect if he does not believe the same thing. Most people, including many well-meaning Christians, believe that modern science has proven that the earth is millions, if not billions, of years old. However, it has not been proven at all.

We studied carbon-14 dating at UVA. Here is how it works:

Carbon is present is all living and formerly living matter. Normal carbon atoms have six protons and six neutrons and are called "carbon-12". A very small percentage of all carbon atoms (one in a trillion) has two extra neutrons and is called carbon-14. These carbon-14 atoms are radioactive and their radioactive emissions can be measured. They are also unstable and over time, collapse into nitrogen, an element with seven protons. [51] Therefore, the more radioactive carbon present in the carbon matter, the younger it is. By using short periods as a standard, it has been proposed that the amount of carbon-14 in organic matter decreases by one-half every 5,730 years, and then over the next 5,730 years by one-half again, and so forth, until all carbon-14 is gone. Of course, no one has actually measured anything for even the first

---

[51] One of the neutrons converts into a proton, giving the atom 7 protons, which make it a different element, nitrogen.

5,730 year period, so it just a proposition, but it is still routinely accepted as a reliable method of dating in scientific circles.

Carbon dating is dependent upon numerous variables, highly technical equipment and is not an exact method by any means. At UVA, our professor volunteered that "you could date the same matter four times and get four different dates, possibly millions of years apart". That is an exact quote from my professor, as best as I can remember. I also remember then asking the professor this question, "Then how do you pick which date to use?" He answered me, "You just pick the one that is most realistic." Yet, despite the enormous problems with the testing, many scientists and all evolutionists unabashedly label anything they want with dates in the millions of years, when in truth they could just as easily date them in the thousands of years.

Because carbon-14 atoms theoretically reduce by half every 5,730 years, the carbon-14 isotope is no longer detectable after 80,000 years. Therefore, the method cannot be used to date anything older than 80,000 years, and anything containing any trace of carbon-14 must be younger than 80,000 years. Yet old earth advocates unabashedly throw around ages in the hundreds of millions of years whenever it suits their purposes.

Carbon-14 dating relies completely upon the assumption that the quantity of carbon matter on the earth has remained constant throughout the earth's entire history. This assumption is an essential ingredient in the carbon dating formula. However, if the earth's quantity of carbon-12 has *not* been constant throughout its history, then the whole carbon dating formula would be fatally flawed. For instance, if the past included a period when there was more plant life than today, as many believe, then the amount of carbon has not been constant throughout the earth's history and the formula is dramatically, if not fatally, flawed. The true age of matter tested would be much younger, perhaps as much as 500 times younger. The irony is that the same people who support carbon-14 dating so vehemently are the same people who also believe that the earth went through periods, such as the Jurassic Period, where there was a considerably higher quantity of vegetation than there is today.

> "It is possible (and, given the Flood, probable) that materials which give radiocarbon dates of tens of thousands of radiocarbon years could have true ages of many fewer calendar years." Gerald Aardsman, - Ph.D., Physicist and C-14 Dating Specialist

This discrepancy led eight scientists to start a research project in 1997 to study the age of the earth. The group was called the RATE group (Radioisotopes and the Age of The Earth). The project lasted eight years. One thing they did was to take coal samples from 10 different coalfields across America and send them to well known dating laboratories. All "old earth" advocates agree that coal is one of the oldest things on earth. The results: All the labs found traces of carbon-14 in every sample. That means that either the carbon-14 method is flawed or all the coal is younger than 80,000 years. Either way, it presented a real problem for all old earth advocates.

The group also used carbon-14 dating to calculate the age of the earth by including in the formula the possibility that there was a time in the earth's history when the earth contained a much higher quantity of vegetation. With this single variation in the carbon-14 formula, they calculated the age of the earth to be 6,000 years old. Imagine that!

Since there is considerable evidence, both from the Bible and from the earth itself, that there was a period in the earth's past when vegetation was more plentiful than it is today, then this formula should be considered more reliable than the one normally used. Nevertheless, any test that threatened the widely held belief that the earth is very old was dismissed out of hand.

The same RATE group also used every available method for calculating the age of inert matter. The results were conclusive. Science simply does not have any reliable means to test the age of anything inert, such as rocks.[52]

No person or machine can examine a rock and determine its age without knowing the form a "young" rock would take and the different forms the same rock would take as it ages over time. None of this information is available today. At this point in time, there is no on-going aging process to observe or use as a standard. It is virtually impossible for anyone

---

[52] The New Answers Book, by Ken Ham Gen. Ed., Master Books, 2006, "Chapter 7 Doesn't Carbon-14 Dating Disprove the Bible?" By Mike Riddle

to say with certainty that he or she knows what form anything took millions of years ago. How could he? A rock is a rock and may have always been a rock. To postulate that it formerly took a different form is pure speculation. It is impossible to know.

If God created the heavens and the earth, then wouldn't He have created them in their finished state? To pick up a finished product and declare that it is very old is to ignore the possibility that it was created in that very form.

Chapter 12
Dinosaurs

Many people think dinosaurs are evidence of a prehistoric world and a very old earth, so I will say just a word about dinosaurs, or as they were formerly called, "dragons".

Recently, 10,000 paintings were discovered in a cave in Peru's Amazon jungle. Here are three of them:

Peruvian cave painting

Peruvian cave paintings

This is nothing new. There are dragon stories and pictures of dragons from China, Africa, India, ancient Rome, the Middle East and Europe.

Babylon had a story of a man named Gilgamesh who killed a huge reptile-like animal.

The city of Nerluc in France was renamed in honor of a dragon that was killed there. It was described as bigger than an ox and with long, sharp, pointed horns on its head.

The story of the knight going out to slay the dragon is well-known.

A 16th century science book, <u>Historia Animalium</u>, has descriptions of small dragons.

Italian and Irish records have detailed descriptions of animals that are obviously dinosaurs.

All of these records are as credible as any other ancient writing and the pictures of the dragons look just like dinosaurs.

A dead plesiosaur dinosaur was caught in the net of a Japanese fishing boat off New Zealand in 1977. It was 32 feet long and weighed 4,000 pounds. It was photographed and examined by the National Science Museum of Japan and chosen as the scientific discovery of the year. Here is a photo.

photo of dead plesiosaur in 1977

The scientific and historical evidence is overwhelming that dinosaurs existed throughout human history.

If I may be permitted to use the Bible as an historical book, in the book of Job, chapter 41, there is the detailed description of the "Leviathan", a giant sea creature with scales like double mail, fearsome teeth and breathed out sparks of fire and smoke. Here is the Bible text:

"Can you draw out Leviathan with a fishhook, or press down his tongue with a cord?

Can you put a rope in his nose, or pierce his jaw with a hook? ...

Will you play with him as with a bird, or will you put him on leash for your maidens?...

Can you fill his skin with harpoons, or his head with fishing spears?

Lay hands on him; think of the battle; you will not do it again!...

No one is so fierce that he dares to stir him up....

I will not keep silence concerning his limbs, or his mighty strength, or his goodly frame.

Who can strip off his outer garment?

Who can penetrate his double coat of mail?

Who can open the doors of his face?

Round about his teeth is terror.

His back is made of rows of shields, shut up closely as with a seal.

One is so near to another that no air can come between them.

They are joined one to another; they clasp each other and cannot be separated.

His sneezings flash forth light, and his eyes are like the eyelids of the dawn.

Out of his mouth go flaming torches; sparks of fire leap forth.

Out of his nostrils comes forth smoke, as from a boiling pot and burning rushes.

His breath kindles coals, and a flame comes forth from his mouth.

In his neck abides strength, and terror dances before him.

The folds of his flesh cleave together, firmly cast upon him and immovable.

His heart is hard as a stone, hard as the nether millstone.

When he raises himself up the mighty are afraid; at the crashing they are beside themselves.

Though the sword reaches him, it does not avail; nor the spear, the dart, or the javelin.

He counts iron as straw, the bronze as rotten wood.

The arrow cannot make him flee; for him slingstones are turned to stubble.

Clubs are counted as stubble; he laughs at the rattle of javelins.

His underparts are like sharp potsherds; he spreads himself like threshing sledge on the mire.

He makes the deep boil like a pot; he makes the sea like a pot of ointment.

Behind him he leaves a shining wake; one would think the deep to be hoary.

Upon earth there is not his like, a creature without fear;" Job 41:1-34

Therefore, it is an historical fact that dinosaurs lived among us as recently as 1977. It is a fact. My personal opinion is that they went the way of the Wooly Mammoth and were hunted to extinction, and, being reptiles, were vulnerable to cold weather.

Chapter 13
More Problems

Humans and Evolution

With evolution, success and dominance are the only criteria that determine what accidental mutations are perpetuated and what are discarded, so why is the highest form of life so different from the "survival of the fittest" stereotype? If winning is everything and all other results are cast onto earth's ruthless trash heap, then how did man evolve?

With this essential part of evolution in mind, consider the human being. He is dexterous, mobile and smart. All of that is very good. But he is also subject to a very strong influence called a conscience. This is the red flag that goes up when we consider behavior that is harmful to others. In almost every case, the conscience seeks to restrain self-centered behavior in favor of the well being of another. It is not hard to see that its influence is stronger in some people than it is in others; nevertheless it is an innate attribute of all humans. If humans were the sole result of evolution, then such would not be the case. There is no place in evolution for such an impractical restraint.

Two Sexes

Consider the enormous difficulties presented by thousands of species evolving step by step, through thousands of steps, two sexes at a time, if with each step, each sex must change in a manner that is totally compatible with the opposite sex. They must both change at exactly the same time in exactly same part of their anatomy to continue the ability to reproduce.

Consider also the differences and complexities of the male and female anatomies of animals and plants. Most of us are somewhat familiar with the sexual anatomy of male and female humans and know how very different they are. Think about the differences right now. For evolution to be possible, every detail of the male sexual anatomy must have come into being through thousands of individual steps that coincided precisely with each of the thousands of individual steps necessary to create the female sexual anatomy so that the two sexes were totally compatible and able to reproduce at each of the thousands of steps along the way. It should be difficult for any rational mind to imagine this happening even once, not to mention thousands of times in each of the thousands of species.

Love

The same can be said of love. Only the most die-hard evolutionist would be so brazen as to say that love is not a powerful force in humans and that we are not constantly doing ridiculous things because of love. The honest among us will acknowledge that love is perhaps the most powerful force on earth and it makes us do crazy stuff. Have you ever been in love? If so, you know what I mean. Where does such an emotion fit into the theory of evolution? It doesn't. It can't. It is a contradiction.

A religious nature

The same can also be said about man's tendency to seek communion with God. Throughout recorded history, humans have always been very religious. From the dawn of time, humans have sought and worshiped some god. Those who do not are in the very small minority. We believe in God, and always have. It is the deepest part of our being. It is not a physical attribute. It is an inner longing. It is part of us. But just like love and the conscience, our religious inclination does not fit anyplace in the evolutionary formula. This aspect of human nature could not be the result of gene mutations that produce a more successful life form. Being religious would not help us progress from one ruthless level of success to the next. There is no place for religious longings in the cold, hard world of natural selection.

Emotions and Art

Why is the highest form of life so emotional and emotionally vulnerable? Being saddled with all these emotions is not consistent with survival of the fittest.

And what about our artistic nature? What possible need would an ape have for painting, poetry, ballet, writing, philosophy and music in his move up?

Mind-Blowing

At a meeting of Psychiatrists, the moderator asked the doctors about the "evolutionary context" of their field. A doctor immediately replied, "Evolution is not a consideration." In other words, when discussing the complexities of the human mind, evolution is out of the question.

The human brain is so extraordinarily complex that doctors have only begun to scratch the surface of its operation. The brain is able to think in a continuous stream, one thought at a time, without overlap. It is able to reason, imagine, question and remember. It is, in fact, beyond our human ability to fully comprehend. To most who devote their energies to the study of the human mind, it is absurd to consider the notion that anything so complex happened by accident.

Consider the ramifications. We humans try to understand the mind with our minds, but we cannot. Do you see the problem? How can that be? How can it be that we cannot fathom the complexities of our own mind with our own mind? What force drove the development of such complexity so far beyond the understanding of the creature in which it resides?

Flying Tigers?

And why are humans so hairless, toothless and clawless? Shouldn't we be like flying tigers if we are the pinnacle of evolution? But as it is, the highest and best form of life today is an unarmed, hairless, smooth-toothed creature that is restrained by a conscience, overcome by love and prone to paint, write poetry and watch operas.

White Egret Orchid

Dancing Girls Orchid

Chapter 14
Bitter Fruit

Evolution has not been a harmless sideshow at the science fair. Some really evil people have taken this ball and run with it. In the hands of killers like Hitler, Stalin and Mao Zedong, evolution was directly behind the deaths of an estimated 200 million people in the 20th Century. Each of these men, and others like them, used this theory as a license to kill to an extent previously unknown in human history.[53]

Karl Marx asked Charles Darwin if he could dedicate Das Kapital to him, but Darwin declined. But really all these men did was take the theory to its logical conclusion, namely that people are 'just animals', that we all just sprang up out of the dirt like all the other animals to scratch our way through life in a ruthless and godless world where it's 'the survival of the fittest' and 'kill or be killed'. To them, there are no eternal consequences whatsoever for their actions. There is no God, no judgment, no heaven and no hell. We just die and rot like a squirrel. So why not make the most of it?

Hitler's Germany was devoutly Lutheran and throughout his political life, he systematically advanced both programs of appeasement and suppression of their faith at the same time. He made numerous references to "God", but his actions were anything but God-fearing. While suppressing the church, Hitler formed the Christian Social Party as a Nazi substitute and stamped out the church. All who

---

[53] Evolution's Fatal Fruit by Tom DeRosa.

protested were imprisoned and killed, the most famous of whom was Dietrich Bonhoeffer. The book Holocaust Forgotten documents Hitler's extermination of five million non-Jews, many for their Christian opposition to Hitler.[54] The 50 million killed in the European theater of WWII can rightfully be laid at his feet.

Stalin, a fanatical atheist, said, "One death is a tragedy. A million deaths is a statistic."[55] He was true to his words and his rampant paranoia killed an estimated 50 million Russians.[56]

Mao Zedong killed an estimated 78 million of his own Chinese people, 45 million in the four years of his Cultural Revolution.[57]

History is proof that these men made the 20th Century the deadliest of all. But so what, people are only animals.

---

[54] by Terese Pencak Schwartz
[55] http://www.brainyquote.com/quotes/quotes/j/josephstal137476.html
[56] International Business Times, Mar 5, 2013 - Dyadkin estimated that the USSR suffered 56 to 62 million "unnatural deaths" during that period, with 34 to 49 million directly linked to Stalin. In "Europe A History," British historian Norman Davies counted 50 million killed between 1924-53, excluding wartime casualties.
[57] A book, based on meticulous research in the Chinese archives, says Mao Zedong was responsible for the deaths of 45 million people between 1958 and 1962 alone. It's called Mao's Great Famine: A History of China's Most Devastating Catastrophe, and its author is the scholar Frank Dikotter, of Hong Kong University.

Chapter 15
The Greatest Irony

A neighbor once told me, "I don't believe in God, I am a scientist."

This attitude is widespread in the biology, anthropology and paleontology communities. Many believe that science and God are irreconcilable. However, this has not always been the case. In fact, history reveals a very different reality. In his book What if Jesus had Never Been Born?, Dr. D. James Kennedy writes about the impact that Jesus and Christians have had on civilization, including science.[58] It was a real eye opener. The comparison between human thought before Christ and after Christ is truly phenomenal. But of interest now is the contribution that Christians have made to science since the Protestant Reformation, because before then, there was no science as we now think of it.

Some books on the history of modern science begin with the year 1500, because "In 1500, science as we know it today was almost unknown, even in Western Europe where it all began" [59] and all significant scientific discoveries were made after that date. The Reformation brought a revolutionary way of thinking to the Christians within the scientific community. Their motto was "Thinking God's thoughts after Him." (Johannes Kepler (1571-1630)) It

---

[58] What if Jesus Had Never Been Born? The Positive impact of Christianity in History, by D. James Kennedy and Jerry Newcombe, 1994, Thomas Nelson Publishers.
[59] Who Discovered What When, David Ellyard, Reed New Holland, 2005, first page of Introduction.

led scientists who studied nature to look for the laws that God set up in nature. It was a totally new and different direction in thinking and led to enormous and numerous scientific breakthroughs. Christians who sought and received God's revelation about nature totally transformed the scientific world. For the most part, they started the development of the scientific world that we know today.

Michael Faraday (1791-1867), the genius who worked with electricity and invented the generator, was a member of a group of scientists who were also Christians. They adhered to the motto "Where the scriptures speak, we speak; where the scriptures are silent, we are silent."

I quote from Dr. Kennedy's book:

"Here is a list of some of the outstanding Bible-believing scientists who *founded* the following branches of science:

Antiseptic surgery, Joseph Lister
Bacteriology, Louis Pasteur
Calculus, Isaac Newton
Celestial Mechanics, Johannes Kepler
Chemistry, Robert Boyle
Comparative Anatomy, Georges Cuvier
Computer Science, Charles Babbage
Dimensional Analysis, Lord Rayleigh
Dynamics, Isaac Newton
Electronics, John Ambrose Fleming
Electromagnetics, Michael Faraday
Energetics, Lord Kelvin
Entomology of living insects, Henri Fabre
Field Theory, Michael Faraday

Fluid Mechanics, George Stokes
Galactic Astronomy, Sir William Herschel
Gas Dynamics, Robert Boyle
Genetics, Gregor Mendel
Glacial Geology, Louis Agassiz
Gynecology, James Simpson
Hydrography, Matthew F. Maury
Hydrostatics, Blaise Pascal
Ichthyology, Louis Agassiz
Isotopic Chemistry, William Ramsey
Model Analysis, Lord Rayleigh
Natural History, John Ray
Non-Euclidean Geometry, Bernard Riemann
Oceanography, Matthew F. Maury
Optical Mineralogy, David Brewster
And on it goes. All of these founders were Bible believers and believers in creation. Creationists are not scientists? Creationists invented science." (pg. 101-102)

I point out that these men *founded* all of these scientific fields. Dr. Kennedy is so right when he says, "Creationists created science". There would be no bacteriology without Louis Pasteur and no oceanography without Matthew Fontaine Maury and no chemistry without Robert Boyle. Each of these men sought and received a revelation from God that opened up an understanding of nature that before then did not exist. Yet many of today's scientists choose to ignore their roots. So now we can add "historically challenged" to the list of evolutionists' shortcomings.

On the wall of every chemistry classroom in the world is the periodic table with all 103 elements and their atomic weight. Here is how that happened. In 1869, a Russian Christian and chemist named Dmitri Ivanovich Mendeleev had a dream. In his dream, he saw the periodic table of elements. He jumped out of bed and drew it. He called it the periodic law. It took him until 1886 to complete his verification of all the elements, but he persevered, because he had before him the picture that he had seen in his dream. His dream is with us today: the periodic table.

Mendeleev also discovered the phenomenon of critical temperature, the temperature at which a gas or vapor may be liquefied by pressure. He was one of many Christians who sought the Lord for insight into his field of study.

Albert Einstein said:

"Science without religion is lame. Religion without science is blind."

"My religion consists of a humble admiration of the illimitable superior spirit who reveals himself in the slight details we are able to perceive with our frail and feeble mind."

"I want to know God's thoughts; the rest are details."

Chapter 16
God and Evolution

Antagonism toward divine creation has played an essential role in evolution's remarkable success. You see, it is this antagonism that has propelled the theory of evolution more than anything else. Some people will avoid accountability to God at any cost. This animosity and this alone, has made the whole ludicrous theory possible and allowed it to endure and become what it is today. There is no doubt whatsoever that if this theory did not have the profound theological implications that it does, that it would have been laughed out of science long ago.

There is an abundance of scientific evidence that evolution is impossible, but Darwinists will never admit it because their real motives totally cloud their scientific judgment. They know about the mystery of the atom and their minds are blown apart when they even try to think about how such a thing came into being. They know genetics and they know that a mutation would just die and could not be passed on, but that does not stop them from talking about it happening many trillions of times. They know there is not one single example anywhere of a significant beneficial gene mutation being passed on to offspring. They know that there is not a single fossil that supports their theory, but they still set out every day with renewed vigor to be the first to find it.

But their thoughts do not linger on such unpleasantries for long, because fast on the heels of these inconvenient thoughts is the inescapable question, "What is the alternative?" That is the real question. In fact, to them, that is the only question that matters, not science. Destroying divine creation is the *only* thing that matters. All else is totally irrelevant.

Evolutionists are not really interested in any analysis of their theory at any level. They really don't care. They don't care whether there is any evidence that supports it, or how it could happen, or all the evidence amassed against it. For there is no amount of evidence that would make any difference whatsoever to them. None of that matters at all. The one and only thing that matters is that they have an alternative to God. That's all. The astonishing persistence of this theory despite its innumerable shortcomings can only be explained by the willingness of its advocates to believe anything at all, no matter how ludicrous, as an alternative to belief in God.

The Bible explains that their motive is to elude accountability. They simply do not want to be accountable to God, so they deny his creation. St. Paul explained it like this:

"For the wrath of God is revealed from heaven against all ungodliness and wickedness of men who by their wickedness suppress the truth. For what can be known about God is plain to them, because God has shown it to them. Ever since the creation of the world his invisible nature, namely, his eternal power and deity, has been clearly perceived in the things that have been made. So they are without excuse; for although they knew God they did not honor him as God or give thanks to him, but they became futile in their thinking and their senseless minds were darkened. Claiming to be wise, they became fools." *Rom. 1:18-22*

That pretty much sums them up. Their senseless minds have been darkened and claiming to be wise, they became fools.

In his book Darwin on Trial, Phillip E. Johnson examines in detail every single premise proposed by Darwinists to support their theory. It is a very scholarly book and any attempt by me to summarize it could not begin to do him justice, nevertheless I will still summarize it as follows: Darwinists propose something, there is no evidence to support it, it violates all that we know about science, they are challenged to show how it could possibly happen, but they cannot do that, so they say, "Evolution is a fact, therefore it had to happen, because what is the alternative?"

Whenever cornered by evidence or scientific impossibilities, Darwin's bulldog, Huxley, never answered the question but always resorted to this simple reply, "What is your alternative?" He found it very effective then, and it is equally effective now. It is their fallback and the very heart of their feelings on the subject. The logical alternative to evolution is so difficult for them to accept that all science and logic are thrown out for the sake of their personal preference which has a lot more to do with freedom of will than science.

It even worked with the Supreme Court of the United States in 1987. When Louisiana wanted to allow its schools to include scientific evidence *against* evolution together with evidence *for* evolution, the court held that the only plausible alternative was creation, and that divine creation cannot be taught. [60] So evolution was allowed to retain its monopoly in the classroom.

Evolutionists must laugh as they say to one another, "You can't argue with success." It is so true. It is exactly how they feel and we would do well to take note of it and respond accordingly. If they throw this in our faces, then what prevents me from returning the favor?

When Darwin's book first came out, there was no shortage of eager followers. Few ideas have ever been met with a warmer welcome than this one. The details were, and still are, irrelevant. Just the possibility that there is no God was enough to excite and bring hope to a large segment of the population. These people are absolutely void of all scientific and ethical integrity. They are boldfaced liars and hucksters of the most grievous kind. Their hysterical aversion to the possibility that there may be a God who created them and to whom they may be accountable has so frozen their judgment that they consider anything, no matter how ridiculous, to be justifiable, just so long as it is godless.

I must cry "Hypocrite!!!", for this is indeed the height of hypocrisy. In the same breathe used to

---

[60] Aguillard v. Edwards, 482 U.S. 578 (1987) Justice Brennan wrote that the purpose of the statute "was clearly to advance the religious viewpoint that a supernatural being created mankind."

In his dissenting opinion, Justice Scalia pointed out that "The people of Louisiana, including those who are Christian fundamentalists, are quite entitled, as a secular matter, to have whatever scientific evidence there may be against evolution presented in their schools, just as Mr. Scopes was entitled to present whatever scientific evidence there was for it."

ridicule divine creation, they use divine creation as the principal reason to support their theory. Their use of "creation" has "created" a scientific monopoly that is as unfair and unscientific as the "superstitions" they so vehemently detest.

They act as though their decision on the matter will have some impact on reality. They seem to believe that their devotion to the godless cause will actually determine whether or not God exists. They push their theory as though the existence of God is dependent upon them. They apparently believe that if they believe evolution strongly enough, then it is impossible for God to exist. Isn't that the dumbest thing you ever heard? While in truth, no amount of blindness to the facts will have any affect whatsoever on the existence of God. This is self delusion of the most grievous kind. They have placed prejudice over science, so they can do what? Convince the world that there is no God. That is evil in its purest form. They have shown utter contempt for you and me, so I will not hesitate to show my contempt for them, and the fact that the only reasonable alternative may be that there is a God with enormous genius, does not deter me a bit.

Consider the true ramifications of their actions. They want to use their scientific credibility to knowingly perpetuate a lie in order to lead the entire world to believe that there is no God. Is this not the only reasonable conclusion that I might draw? Yes, to me it is. They are not only liars; they want to use their lie to condemn the whole world. Think of the magnitude of their evil. But their evil does not stop there. They also want to deny God the rightful glory He is due for His wondrous creation.

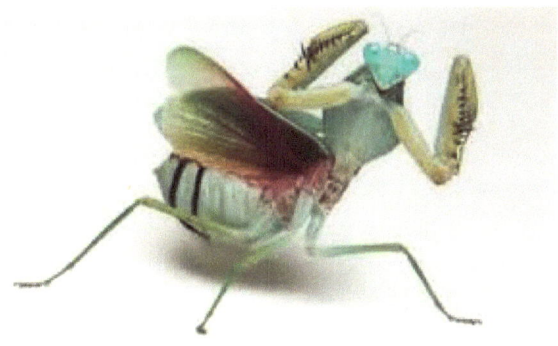

Chapter 17
The Bible to the Rescue

As each successive generation applauds its own particular genius, it should look over its shoulder and consider the past generations that applauded themselves with equal gusto, and consider that the next generation will feel the same about them. Future generations of scientists will look back on us as hopelessly uninformed. At one time, science believed that the earth was flat and that the sun revolved around it.[61] An apple had to fall on Isaac Newton's head before anyone thought of something as basic as gravity. Louis Pasteur faced fierce opposition from the entire French medical community over his rabies vaccine. George Washington died because doctors thought Washington needed to get rid of more blood. The entire field of molecular biology did not even start until 1953 when Watson and Crick discovered the DNA molecule. Yet science remains imminently impressed with itself. In reality, science has only scratched the surface in its quest to unravel the mysteries of nature, and would do well to acknowledge its many mistakes, countless unanswer-ed questions, and not be so ridiculously biased about the biblical account of creation. I remind the reader that it was the Protestant Reformation and the divine inspiration received by scientists within its borders that FOUNDED over 29 different fields of science. Before these men pursued divinely inspired science,

---

[61] In 1492, most of Christopher Columbus' crew feared they would fall off the edge of the earth.

the world wallowed in ignorance and superstition. Numerous natural conditions lie right under everyone's noses that are easily explained by the biblical account, yet science prefers pure speculation and outlandish imagination to anything biblical.

Some Christians feel compelled to adopt a compromise position such as "God used evolution" or "life evolved as prompted by intelligent design". Such compromise is not necessary. The Bible does not need our help. Just as we benefited greatly from some basic science, we will also benefit greatly from some basic Bible. I find it fascinating and I hope you will too.

In the Beginning

In college, I learned about the basic components of the universe: time, space and matter. These words mean just what you think they mean: time, space and matter. With that in mind, let's look at the first things God created.

"In the beginning" *Gen.1:1*

Before creating the universe, God dwelt in eternity, a realm without time. There was no need for time. There was no beginning, and no end. So the first thing God created when He created our world was to create time. Therefore the first words of the Bible are **"In the beginning"**, denoting the beginning of time. God created "time".

"In the beginning, God created the heavens" *Gen.1:1*

When God created the heavens, He created space. Now, there was time and space.

"In the beginning, God created the heavens and the earth." *Gen.1:1*

Then God created the earth which is "matter".

So, the first sentence of the Bible starts off with the creation of the three basic components of the universe: time, space and matter.

Energy

We also learned in college about the three basic forms of energy: nuclear, photoelectric and gravitational.

When God created matter, He created the nucleus of atoms, which are held together with nuclear energy, so God created nuclear energy.

When God said "Let there be light", He created photoelectric energy.

When God formed the earth into a large solid mass, God created gravitational energy.

So in the very beginning of creation, God created the three forms of energy.

Basic science in the first couple of verses.

Waters

According to the biblical account, God created the heavens and the earth in a surprising fashion. First, He placed the firmaments (heavens) in the midst of waters. In so doing, He divided the waters into two separate locations, with the heavens between them.

Waters

....

Heavens

....

Waters

A curious start, isn't it? We will soon see why. Then God gathered together the waters that were under the heavens into one place together with the land, and dry land appeared. He formed the earth with half of the water and left the other half *above* the heavens as a layer of thick clouds that enclosed the entire earth. The earth was totally surrounded by a canopy of water, just above the sky. *Gen. 1:1-9.*

Waters (in form of clouds)

....

Heavens

....

Earth with water

From the time of creation until Noah's Flood, there was no rain on the earth. Instead, the earth was watered by a mist that came up from the earth that watered the whole face of the ground. *Gen 2:5-6.*

> "for the Lord had not caused it to rain upon the earth,… but there went up a mist from the earth, and watered the whole face of the ground." *Gen. 2:5-5*

This mist, together with the thick canopy of clouds around the earth, created a greenhouse effect for the whole earth. The sun's harsh rays were largely blocked out by the clouds and, inside the canopy, the earth's moisture and heat were held inside by the canopy of water around it. It was warm, wet and mild. This produced an ideal environment for life. It was lush and green, with an abundance of plant and animal life. This environment lasted for between 1,500 and 2,000 years, until Noah's Flood.

As a prelude to Noah's Flood, the Bible says:

> "Then the Lord saw that the wickedness of man was great in the earth, and that every intent of the thoughts of his heart was only evil continually." *Gen. 6:5*

So God destroyed every living thing on the face of the earth with a flood, except of course Noah and those with him in the ark. *(Gen. 7:21)* All the water for Noah's Flood came from the thick cloud cover that surrounded the earth and from fountains that burst up from within the earth. It required a lot of water to cover the whole earth and all its mountains. The earth's current cloud cover does not have the capacity to rain continually for forty days and forty nights. But the pre-flood canopy did.

> "all the fountains of the great deep were broken up, and the windows of heaven were opened." *Gen. 7:11*

This flood covered all the earth, including the highest mountains, and only Noah and those with him in the ark survived. The waters prevailed upon the earth for 150 days.

After this flood, the earth was a totally different place. Instead of being surrounded by a thick canopy of water, the earth was now open to the sun. Instead of a greenhouse-like environment, the earth was now subject to harsh weather, radically different climates and dramatic seasonal changes. Instead of being watered by a mist, the earth was now watered by rain via the current water cycle. Instead of having a predominance of lush green vegetation, there was now a wide variety. Instead of having an unsculptured landscape, it now had sharp precipices and gorges.

Different Climates

Geologists believe that the earth has gone through dramatic changes in its climate. They believe that long ago, the earth was very warm and conducive to lush vegetation. Then, for some reason, its climate abruptly changed and it got much colder. They really don't have any idea how or why this might have happened. So, they have come up with some pretty bizarre speculations; for instance a giant asteroid hitting the earth. However, such an event would have had only a temporary effect. It does not explain all the changes that the biblical flood explains.

Neanderthal and Cro-Magnon Man

Many remains of early man have been unearthed. Their physical characteristics were superior to ours and they are universally acknowledged as human. He was a beautiful specimen. There is certainly no evidence that he was inferior. His brain space was much larger than ours and he is bigger and stronger than us, so why did he come *before* us? And why is he always drawn to look like such an idiot?

I submit that both Neanderthal and Cro-Magnon man were pre-flood people who enjoyed the near perfect environment afforded them. They were the first created and had not yet succumbed to the harsh weather, bad food and disease that plagued the post-flood generations.

Noah's Flood

The earth's crust has layers. They are most evident when one is driving on a road that has been cut through a hill or mountain. Often, one can see that these layers are slanted or tilted, sometimes in an extreme fashion. So, the next step is to consider how it happened. Some force caused the layers of the earth's crust to tilt. Something pushed them up.

Another curious fact about the earth's crust is that the fossils of marine plants and animals can be found thousands of miles from the ocean; some of them even on the top of mountains.

Another look at the crust reveals deep valleys and gorges that have been cut by the massive flow of water.

It does not suffice to say "It happened", or "The earth is the result of cataclysmic events". That really does not explain it. Geology describes what it observes, but does not explain how it came to be.

But all of this is easily explained by the biblical flood; the tilt in the layers of the crust ("all the fountains of the great deep were broken up"), marine fossils on mountaintops ("all the high mountains under the whole heaven were covered"), deep gorges ("the waters prevailed and increased greatly upon the earth") and so much more. *Gen. 7:11, 18-19*

Fossils

In the normal cycle of life, plants and animals die and decompose where they fall down, usually on the surface of the earth.

I have seen many a dead animal, and know very well that they immediately begin to rot. After only a few days, the process of decay is well advanced. After only a week or two, there is nothing left but bones. The only exception is when it is very cold.

In my law practice, I was involved in the moving of a small family cemetery. It had been there for about 100 years. I spoke with the man who performed the task and he told me that the best he could do was scoop up some dirt that looked darker than the dirt around it, because there was no physical remains of any kind in the graves. Even the bones had completely decomposed. That is what normally happens to all plants and animals after they die. They rot completely. They do not form fossils.

Fossils are formed only when there is a precise sequence of events. First, it must happen fast, very fast. The entire process, from start to finish, must finish before the plant or animal has decomposed, and as we just said, that happens fast.

Second, the dead plant or animal must be completely covered by soft wet dirt (mud) before it begins to rot.

Third, the mud with the plant or animal inside of it must be subjected to sufficient pressure to compress the mud into solid rock.

If all three of these factors are present, then a rock cast of the plant or animal will be formed. If all three are not present, the plant or animal will decompose and disappear and no fossil will be formed. Obviously, the events necessary to form a fossil are not every day events in nature.

Fossils are not found throughout the earth's crust. They are usually only found within one thin layer within the crust. This indicates that their formation was largely the result of a one-time event.

Now consider what happened in Noah's Flood. All life was drowned and covered by enough water to cover the surface of the entire earth, even the top of the mountains. Millions of dead animals sank to the bottom, where some became covered in mud. Then this muddy bottom was immediately covered with trillions of tons of water, enough to compress mud into solid rock. Then after several months, the waters began to evaporate and formed the earth's new cloud cover. This left behind hard evidence of what had just happened. Evidence that could not have been formed any other way. The fossil record is available only because of the flood.

"We have to admit that there is nothing in the geological records that runs contrary to the views of conservative creationists." Edmund Ambrose, -Evolutionist

Crust

First of all, it is evident that the earth was formed by a tremendous force that compressed it into its spherical shape and density. How else could it be so round and tightly packed? But such compacting forces are not natural in nature. Left to themselves, things don't just constrict into spheres.

"He drew a circular horizon on the face of the waters." *Job 26:10*

As previously covered, there is evidence that a force of some kind pushed the earth's crust upward, forming mountains and tilting the crust.

All of this could be easily explained by the hand of God forming the earth initially and the fountains of water erupting through the crust at the time of Noah's Flood. But geologists much prefer absurd speculation over the Bible. Have we not all seen how absurd they can be? Consider the imaginative pictures showing South America and Africa fitting nicely together and then South America just floating three thousand miles across the ocean to where it is now. How crazy is that? Can an entire continent move and leave nothing in its wake?

Noah's flood explains it all. It explains the fossils, the presence of fossils deep inland and on mountaintops, the disruptions of the crust and the deep gorges. It supplies simple answers to difficult questions, yet it is off-limits because it is biblical. What a shame.

Human Sexuality

With all other animals, the two sexes are almost identical in terms of instincts, behavior, emotions and all other non-physical traits. It makes little difference whether a dog, cat, horse or squirrel is male or female. They all act and think the same. Such is certainly not the case with humans. We are absolutely unique in this regard.

Every human who has not lived in solitary confinement all of his life has wondered about the mystery of the opposite sex. To most women, men are a total mystery, and to most men, women are a total mystery. Evolution certainly cannot provide even the most strained attempt to explain this difference in our species and to my knowledge, evolutionists have not even tried.

But I submit that the Bible can.

When God created Adam, the Bible says a curious thing about him:

> "So God created man in His own image; in the image of God He created him; male and female He created them." *Gen. 1:27*

This is curious because this statement comes before Eve was created.

Then to create Eve, God took a rib out of Adam and

> "the rib which the Lord God had taken from man He made into a woman, and He brought her to the man." *Gen. 2:22*

Now let's put the two facts together. First, when God created Adam, "in the image of God He created him; male and female He created them." Adam was both in the image of God and both male and female. Adam was created in the full image of God, whose image must of necessity include both the male and female nature. There is much about this fact that I do not understand, but we will continue anyway.

Second, to create Eve, God took a rib out of Adam. He took it out of him. He took something out of the man. Now Adam, the man, was missing something that he formerly had. And from that which he no longer had, God created woman.

From this we can understand some things that have puzzled mankind since the beginning of time. Men have a hard time fully understanding women because they don't have something that women have. What women have, men no longer have. *They don't have it.* It was taken out of Adam. And vice versa. Women have a hard time fully understanding men because they don't have what remained in Adam. *They don't have it.* God, in a manner of speaking, split Adam into two parts, each with some of the total traits that make up the image of God.

On the other hand, God has it all. God has no problem relating to both men and women, and both men and women have no problem relating to God.

I am not suggesting that God is a woman or that Adam had any female physical traits. I am only suggesting that what God took out of Adam to create Eve, Adam no longer had and that explains a lot. And it explains why the two sexes compliment each other so nicely, and it explains the mystery of marriage, where two can join together and become one flesh.

This difference between human sexuality and all other species is just so vast that it is ludicrous to propose that we are just the next link up from the closest form of life, whatever creature that may be.

The Races

Evolutionists have had no consistent explanation for the different ethnic races. At first, they proposed that some races were inferior to others and this was a proof of evolution.[62] At UVA, our evolution

---

[62] Called "Social Darwinism". A century ago, evolutionist believed that evolution was proven by the inferiority of some races, namely the

professor dodged the question, leading us to believe that evolution had no explanation for the different ethnic races and our textbook was likewise silent on the subject.

However the Bible explains it quite simply when it describes a curious and momentous event at the Tower of Babel. At that time, all people on earth were of one race and one language. But at the Tower of Babel, God gave each family a different language and the families scattered all over the earth.

> "'Come let Us go down and there confuse their language, that they may not understand one another's speech.' So the Lord scattered them abroad from there over the face of all the earth." *Gen. 11:7-8*

Before the Lord scattered each family over the face of the earth, they were all part of a single gene pool. But after they scattered, they were no longer a part of the whole. They took with them their own unique family gene pool and one gene pool was split into hundreds of different gene pools. Some families would have had some blonde hair genes and some none. The same was true for eye color and skin color and hair texture and height and so on. Since each family was separated unto itself, each family was forced to interbreed within its own family. This inbreeding eventually resulted in the sifting out of the

---

Negro race, and progressed up to the superiority of the Caucasian race. Hunter, G.W. 1914. *Hunter's biology*. New York, New York: American Book Company. pages 263–265, quoted in *Inherit the Wind*. Such is the consistent hogwash of their theory.

less plentiful and recessive genes and the domination of the most plentiful and dominant genes within each family. Distinct family traits began to emerge. In a matter of only a few generations, very distinct traits were taking over each family's gene pool. As a result, each different family, with its own different language became a different ethnic group and eventually its own country. All the different ethnic groups have since been grouped into the five different races. They may be separate ethnic families and races, but since they all have the same chromosome structure, they can still interbreed.

This is very different from evolution. It is the truth.

Spinning Around

As we all know, the earth rotates on its axis once every 24 hours. It spins at a rate of 1,000 mph. The earth's weight in tons is 66 with twenty zeros. Rotation like that requires a lot of force. What makes it spin and keep on spinning? What energy source provides the power to spin an object weighing 6,600,000,000,000,000,000,000 tons at 1,000mph all the time without slowing down?

"He hangs the earth on nothing." *Job 7:7*

Jupiter is 1,000 times larger than the earth and it rotates even faster. It has more surface area to spin, yet spins around once every ten hours. What force keeps such a mass spinning at such a rate without a single pause?

The earth travels around the sun at a speed of 66,600mph. It never slows down. It just keeps going at the same speed all the time and never leaves its orbit. What kind of power makes this happen?

Our solar system moves through the Milky Way Galaxy at a speed of 558,000mph. How? What pushes it or pulls it? It is a solar system! It is big, real big. What force can move such a mass as that?

I have spun things. They always slow down after a while and if you want to keep it spinning, you have to apply more force. When I was a child, I had a hula-hoop. It was a large plastic hoop that spun around your waist. But it only spun if you spun it with your own force. It never once spun by itself. If I stopped wiggling my hips, it simply fell to the ground.

The US Patent Office will not even accept applications for a "perpetual motion" machine, because it is impossible for man to make such a thing. Yet, in nature, it is normal. Electrons spin, moons spin and planets spin; all without any apparent outside help. What makes it all happen? Science can easily state that it happens and that it is a natural occurrence in nature. Of course it is. But science cannot really provide the answer to the source of this enormous and constant energy.

Total Harmony

The earth contains several systems operating simultaneously in perfect harmony. Let's look at some.

The Water cycle

The earth contains the perfect amount of water in its clouds and on the earth's surface for the cycle to flow perfectly as necessary to support all life. It never runs out and it is constantly purified in the process.

> "God understands its way, and He knows its place. For He looks to the ends of the earth, and sees under the whole heavens, to establish a weight for the wind, and apportion the waters by measure. When He made a law for the rain, and a path for the thunderbolt." *Job 28:23-26*

Water is an almost magical substance that can take the form of solid, liquid or gas. I marvel at its qualities even when I wash my hands. I watch it flow together, wash away my dirt and then disappear off my hands when it evaporates. It is perfect for what it does, and it does a lot.

The Air cycle

The earth's atmosphere also flows in a wondrous cycle. The plants breathe carbon dioxide and give off oxygen. Animals breathe oxygen and give off carbon dioxide. The air contains the perfect combination of elements (oxygen, carbon dioxide, nitrogen, etc.) as necessary for all life. And the earth contains the perfect amount of plants, both on land and in the oceans, as necessary to produce the oxygen needed to support life. It never runs out for the plants or for the animals. It just keeps going in a perfect cycle.

"to establish a weight for the wind" *Job 28:25*

The Soil cycle

The soil contains all the nutrients needed by plants, but the soil would soon give out if not for its own cycle. The same plants that suck out the nutrients also die and decay in the soil. And the animals that eat the plants also leave waste that is a powerful fertilizer. And when the animal dies, it also rots and fertilizes the soil. So the cycle goes on and the soil is constantly renewed. It never gives out.

Perfect Harmony

Each of the several cycles has a complexity and wonder beyond human understanding and still each of these different systems operates in perfect harmony with all the other systems. It is impossible for these several different systems to operate together in more perfect harmony than they do. Together they form an ideal environment for life at a level of supreme opportunity for its inhabitants.

Behold, it was very good

Astronomers are always searching for other planets in the universe that might contain life. They compiled a list of essential characteristics that is needed for a planet to sustain life. One of the criteria is the distance from a sun. If it is too close, it would be too hot, and if it is too far, it would be too cold. This study showed the distance that our earth must be from the sun in order to provide the right temperature on earth and it is a very, very small window. The earth is directly in the middle of it. The earth is 93 million miles from our sun, but if it happened to be only a few miles different in either direction, then we could not live here.

If our earth was even a tiny bit larger, then its gravity would be too strong for us to get about, and if earth was even a tiny bit smaller, then its gravity would be too weak for normal life.

If earth had a different quantity of water than it has, then there would be either too much or too little for the necessary cycle that began after Noah's flood.

If the earth's atmosphere had different elements than it has or if the proportions of the different elements were different than they are, then there could not be life here.

If the earth's crust did not contain the nutrients that it does, then life could not exist.

If the earth did not have its canopy of clouds and atmosphere, then the suns rays would kill us.

If the moon were not the size and distance that it is, then the tides would not move the oceans and rivers around like it does and they would not get the mix they need.

If the earth did not have a hot inner core, then its surface would be too cold for even the sun to heat it up. But as it is, the earth's core keeps the surface of the earth at just the right temperature (55° F).

The earth's orbit results in a perfect length of day and night and seasons of the year.

It was created perfectly.

The Perfect Balance
God created man and his world with God's relationship with man in mind.

If life were too easy, man would not seek God. And if life were too difficult, man would have no choice but to seek God. As it is, life on earth is just the right balance between providing for us and challenging us. The world created for man is sufficient to take care of man, yet difficult enough to challenge man.

God wanted man to have enough provision to be able to be independent if he chose that path, but vulnerable and needy enough for the enlightened among us to seek Him. God gave man a total freedom of will to choose Him or not. In order to place this freedom of will in the proper environment, God created a world where man is not so needy that he has no choice but to seek God or die. Nor is man in a world where he is so abundantly provided for that he would not feel the need for God.

God made us different than the animals. He made us so that we could relate to Him. God made man in His own image and after His own likeness. That means we are like God in many ways, and since we are like Him, we can relate to Him. We have emotions and feelings like He does. We can think and draw conclusions like He does. We have a conscience and know right from wrong. We love and hate. We can enter into strong relationships and friendships. We feel loyalty. These are all attributes that we share with God. Therefore, we can hear Him speak and understand. God and man can talk to each other in terms that we both share. It is not like a man relating to a dog, where the level of communication is so unequal. We can really relate to God, and God to us. He made us that way.

Chapter 18
Imagine

Imagine the thought that went into inventing so many marvelous species of plants and animals, each one a masterpiece of perfection for the life it leads in its very own niche.

Imagine a solar system where huge planets and suns hang in space.

Imagine the genius that invented gravity, electricity and all the earth's systems.

Imagine creating milk that you can drink or make into butter or cheese, or sour cream or yogurt or cottage cheese or buttermilk or curds and whey.

Imagine inventing a microscopic bit of pollen that can fertilize a flower that produces a tiny seed that grows into a huge tree.

Can you look at a rock or block of wood and imagine that it is really a bunch of atoms with electron spinning around? But if it were not, then wood could not burn, chemicals could not be mixed and useful goods could not be produced.

Imagine atoms combining to form a molecule, that combine with other molecules to form a tiny gene that combine in large numbers to form a tiny chromosome that joins with other chromosomes that swim to a tiny egg and penetrate the egg and join with it and each chromosome finds its counterpart and wraps around it to form a new chromosome that contains all the information for a new living thing to grow up to maturity and produce its own genes.

Imagine a sub-microscopic gene that somehow contains enough information to fill 750 large volumes.

Imagine creating 60,000,000,000 different fingerprints at a time?[63]

Then imagine many of its inhabitants and beneficiaries telling its creator to his face that He had nothing to do with it, but rather that its inhabitants have concluded without any real evidence to support it, that it all just happened by accident without Him.

---

[63] Clark Taylor

Chapter 19
A Spider's Tale

One night I noticed a big spider outside my window. The spider was weaving a new web. I watched. He went round and round very fast laying down new web fibers without making any mistakes. Here he was, a very simple animal, and yet able to lay down a web without pulling the spokes away from their straight lines or touching the strand next to it. In fact, each strand was exactly the same distance from the strand next to it all the way around. It was truly a marvel of nature and I told God so. I told God right then and there that I was very impressed with His creation. I felt a sweet move of the Holy Spirit over me. I could tell that God appreciated someone appreciating His creation. I learned something very important that night.

Evolution hangs over the head of every serious Bible student like a black cloud, because creation is not just mentioned in the book of Genesis, it is mentioned throughout the whole Bible. The biblical references to the "God who created the heavens and the earth" are in the hundreds. Moses, David, Solomon, Peter, Paul and Jesus believed that God created all things and said so. I could set out hundreds of Bible verses from nearly every author who contributed to the Bible. It would be an impressive list of godly men who walked with God and heard from God. Their revelations on life and the universe make the combined so-called knowledge of all Darwinists put together look like simpletons.

Every Sunday, preachers all over the world expound on some verse written by the Apostle Paul, analyzing it up and down and all around, constantly referring to it as the word of God. This same Paul believed and wrote that God created the heavens and the earth.

One of the most widely read books in the Bible is the Gospel of John. Here is how it begins:

> "In the beginning was the Word, and the Word was with God, and the Word was God. He was in the beginning with God. All things were made through Him and without Him nothing was made that was made."

Moses knew that God created the heavens and the earth. Which evolutionist can part the Red Sea or fellowship with God without food or water on top of a mountain for 40 days?

King David knew that God created the heavens and the earth. Which evolutionist killed a ten foot giant when he was a teenager, wrote 100 psalms, and led a nation to greatness?

Paul knew that God created the heavens and the earth. Which evolutionist has seen Jesus in the sky, worked mighty miracles and turned the world upside down?

Jesus knew that He created the heavens and the earth. Which of the foul evolutionists has raised the dead, healed the sick, walked on water, rose from the dead and sits at the right hand of God Almighty?

The fear of God is the beginning of wisdom *(Ps. 111:10, Prov. 9:10)*. Those who consider themselves wise should consider this.

I love the book of the Revelation, not because I understand any of it, because I don't, but because it offers us a rare glimpse into God's realm that no other book does. The Bible would not be the Bible without it. In this book, we see heaven and earth from God's perspective. At one of its most exciting moments, it says:

> "And the four living creatures, each having six wings, were full of eyes around and within. And they do not rest day and night, saying:
>
> Holy, holy, holy

Lord God Almighty,
Who was and is and is to come!
Whenever the living creatures give glory and honor and thanks to Him who sits on the throne, who lives forever and ever, the twenty-four elders fall down before Him who sits on the throne and worship Him who lives forever and ever, and cast their crowns before the throne, saying:

You are worthy, O Lord.
To receive glory and honor and power;
For you created all things,
And by your will they exist and were created." *Rev. 4:8-11*

Addendum
Quotes by Scientists

"No matter how numerous they may be, mutations do not produce any kind of Evolution." **Pierre-Paul Grasse,** - Evolutionist

"The pathetic thing is that we have scientists who are trying to prove Evolution, which no scientist can ever prove." **Dr. Robert Millikan,** - Nobel Prize Winner and Eminent Evolutionist

"Overwhelming strong proofs of intelligent and benevolent design lie around us ... The atheistic idea is so nonsensical that I cannot put it into words." **Lord Kelvin,** - Vict. Inst., 124, p267

"The best physical evidence that the earth is young is the dwindling resource that Evolutionists refuse to admit is dwindling ... the magnetic energy in the field of the earth's dipole magnet ... To deny that it is a dwindling resource is phony science." **Thomas Barnes,** - Ph.D., Physicist

"The likelihood of the formation of life from inanimate matter is one to a number with 40,000 noughts after it ... It is big enough to bury Darwin and the whole theory of Evolution ... if the beginnings of life were not random, they must therefore have been the product of purposeful intelligence." **Sir Fred Hoyle,** - Astronomer, Cosmologist and Mathematician, Cambridge University

"It is easy enough to make up stories, of how one form gave rise to another, and to find reasons why the stages should be favoured by natural selection. But such stories are not part of science, for there is no way of putting them to the test." Darwin's Enigma by **Luther D. Sutherland,** Master Books 1988, p89

"Is it really credible that random processes could have constructed a reality, the smallest element of which - a functional protein or gene - is complex beyond ... anything produced by the intelligence of man?" **Michael Denton,** -Molecular Biologist Evolutionist: A Theory in Crisis (London: Burnett Books, 1985) p 342.

"When I make an incision with my scalpel, I see organs of such intricacy that there simply hasn't been enough time for natural Evolutionary processes to have developed them." **C. Everett Koop, -** Former US Surgeon General

"Modern apes ... seem to have sprung out of nowhere. They have no yesterday, no fossil record. And the true origin of modern humans ... is, if we are to be honest with ourselves, an equally mysterious matter." **Lyall Watson, -** Ph.D., Evolutionist

"Although bacteria are tiny, they display biochemical, structural and behavioral complexities that outstrip scientific description. In keeping with the current microelectronics revolution, it may make more sense to equate their size with sophistication rather than with simplicity ... Without bacteria life on earth could not exist in its present form." **James A. Shipiro, -** "Bacteria As Multicellular Organisms", Scientific America, Vol. 258, No.6 (June 1988)

"That a mindless, purposeless, chance process such as natural selection, acting on the sequels of recombinant DNA or random mutation, most of which are injurious or fatal, could fabricate such complexity and organization as the vertebrate eye, where each component part must carry out its own distinctive task in a harmoniously functioning optical unit, is inconceivable. The absence of transitional forms between the invertebrates retina and that of the vertebrates poses another difficulty. Here there is a great gulf fixed which remains inviolate with no seeming likelihood of ever being bridged. The total picture speaks of intelligent creative design of an infinitely high order." **H. S. Hamilton, -** (MD) "The Retina of The Eye - An Evolutionary Road Block"

The entire hominid collection known today would barely cover a billiard table, but it has spawned a science because it is distinguished by two factors which inflate its apparent relevance far beyond its merits. First, the fossils hint at the ancestry of a supremely self-important animal - ourselves. Secondly, the collection is so tantalizingly incomplete, and the specimens themselves often so fragmented and inconclusive, that more can be said about what is missing than about what is present. Hence the amazing quantity of literature on the subject ever since Darwin's work inspired the notion that fossils linking modern man and extinct ancestor would provide the most convincing" proof of human Evolution, preconceptions have led evidence by the nose in the study of fossil man." **John Reader,** "Whatever Happened to Zinjanthropus?", New Scientist Vol. 89, No.12446 (March 26,1981) pp 802-805)

"The only competing explanation for the order we all see in the biological world is the notion of Special Creation." **Niles Eldridge,** - PhD., Palaeontologist and Evolutionist, American Museum of Natural History

"I have little hesitation in saying that a sickly pall now hangs over the big bang theory." **Sir Fred Hoyle,** -Astronomer, Cosmologist, and Mathematician, Cambridge University

2. Darwin's Own Confession

"To suppose that the eye with all its inimitable contrivances for adjusting the focus to different distances, for admitting different amounts of light and for the correction of spherical and chromatic aberration, could have been formed by natural selection, seems, I Freely Confess, Absurd In The Highest Degree." **Charles Darwin,** - Origin of Species, Chapter Difficulties.

"As yet we have not been able to track the phylogenetic history of a single group of modern plants from its beginning to the present." **Chester A. Arnold,** - Professor of Botany and Curator of Fossil

Plants, University of Michigan, <u>An Introduction to Paleobotany</u> New York: McGraw-Hill, 1947, p.7

"Despite the bright promise that paleontology provides means of 'seeing' Evolution, it has provided some nasty difficulties for Evolutionists, the most notorious of which is the presence of 'gaps' in the fossil record. Evolution requires intermediate forms between species and paleontology does not provide them." **David Kitts,** - Ph.D. "Paleontology and Evolutionary Theory", <u>Evolution,</u> Vol. 28 Sept. 1974 p.467

"Hundreds of scientists who once taught their university students that the bottom line on origins had been figured out and settled are today confessing that they were completely wrong. They've discovered that their previous conclusions, once held so fervently, were based on very fragile evidences and suppositions which have since been refuted by new discoveries. This has necessitated a change in their basic philisophical position on origins. Others are admitting great weaknesses in Evolution theory." **Luther D. Sutherland,** - <u>Darwin's Enigma: Fossils and Other Problems</u>, 4th edition Santee, California: Master Books, 1988 pp.7-8

"Micromutations do occur, but the theory that these alone can account for evolutionary change is either falsified, or else it is an unfalsifiable, hence metaphysical theory. I suppose that nobody will deny that it is a great misfortune if an entire branch of science becomes addicted to a false theory. But this is what has happened in biology: ... I believe that one day the Darwinian myth will be ranked the greatest deceit in the history of science. When this happens many people will pose the question: How did this ever happen?" **S. Lovtrup,** - <u>Darwinism: The Refutation of a Myth,</u> London: Croom Helm, p.422

"If one allows the unquestionably largest experimenter to speak, namely nature, one gets a clear and incontrovertible answer to the question about the significance of mutations for the formation of species and Evolution. They disappear under the competitive conditions of natural selection, as soap bubbles burst in a breeze." **Herbert Nilson,** -Evolutionist, Synthetische Artbildung Lund, Sweden: Verlag CWK Gleerup Press, 1953, p 174

"The uniform, continuous transformation of Hyracotherium into Equus, so dear to the hearts of generations of textbook writers, never happened in nature." **George Simpson,** - Palaeontologist and Evolutionist

3. Fossils

"Evolution requires intermediate forms between species and paleontology does not provide them." **David Kitts**, Palaeontologist and Evolutionist,

"As is well known, most fossil species appear instantaneously in the fossil record." **Tom Kemp,** Oxford University

"The fossil record pertaining to man is still so sparsely known that those who insist on positive declarations can do nothing more than jump from one hazardous surmise to another and hope that the next dramatic discovery does not make them utter fools ... Clearly some refuse to learn from this. As we have seen, there are numerous scientists and popularizers today who have the temerity to tell us that there is 'no doubt' how man originated: if only they had the evidence..." **William R. Fix,** - The Bone Pedlars, New York: Macmillan Publishing Company, 1984, p.150

"The intelligent layman has long suspected **circular reasoning** in the use of rocks to date fossils and fossils to date rocks. The geologist has never bothered to think of a good reply." **J. O' Rourke,** The American Journal of Science

"In most people's minds, fossils and Evolution go hand in hand. In reality, fossils are a great embarrassment to evolutionary theory and offer strong support for the concept of Creation. If Evolution were true, we should find literally millions of fossils that show how one kind of life slowly and gradually changed to another kind of life. But missing links are the trade secret, in a sense, of palaeontology. The point is, the links are still missing. What we really find are gaps that sharpen up the boundaries between kinds. It's those gaps which provide us with the evidence of Creation of separate kinds. As a matter of fact, there are gaps between each of the major kinds of plants and animals. Transition forms are missing by the millions. What we do find are separate and complex kinds, pointing to Creation." **Dr. Gary Parker, -**Biologist/Palaeontologist and Former Ardent Evolutionist

"...I still think that, to the unprejudiced, the fossil record of plants is in favour of special creation. Can you imagine how an orchid, a duckweed and a palm tree have come from the same ancestry, and have we any evidence for this assumption? The Evolutionist must be prepared with an answer, but I think that most would break down before an inquisition." **Dr. Eldred Corner, -** Professor of Botany at Cambridge University, England: Evolution in Contemporary Botanical Thought, Chicago: Quadrangle Books, 1961, p.97)

"Fossils are a great embarrassment to Evolutionary theory and offer strong support for the concept of Creation." **Gary Parker,** PhD, Biologist/Palaeontologist and Former Evolutionist

"So firmly does the modern geologist believe in Evolution up from simple organisms to complex ones over huge time spans, that he is perfectly willing to use the theory of Evolution to prove the theory of Evolution (p.128) ... one is applying the theory of Evolution to prove the correctness of Evolution. For we are assuming that the oldest formations contain only the most primitive and least complex organisms, which is the base assumption of Darwinism ... (p.127) If we now assume that only simple organisms will occur in old formations, we are assuming the basic premise of Darwinism to be correct. To use, therefore, for dating purposes, the assumption that only simple organisms will be present in old formations is to thoroughly beg the whole question. It is arguing in a circle (p.128)." **Arthur E. Wilder-Smith,** - Man's Origin, Man's Destiny, Harold Shaw Publishers, 1968, pp127-128

"It cannot be denied that from a strictly philosophical standpoint, geologists are here arguing in a circle. The succession of organisms has been determined by the study of their remains imbedded in the rocks, and the relative ages of the rocks are determined by the remains of the organisms they contain." **R. H. Rastall,** - Lecturer In Economic Geology, Cambridge University: Encyclopaedia Britannica, Vol.10. Chicago: William Benton, Publisher, 1956, p.168

"I admit that an awful lot of that (fantasy) has gotten into the textbooks as though it were true. For instance, the most famous example still on exhibit downstairs (in the American Museum of Natural History) is the exhibit on horse Evolution prepared fifty years ago. That has been presented as literal truth in textbook after textbook. Now, I think that that is lamentable, particularly because the people who propose these kinds of stories themselves may be aware of the speculative nature of some of the stuff. But by the time it filters down to the textbooks, we've got science as truth and we have a problem." **Dr. Niles Eldredge,** - Palaeontologist and Evolutionist

4. DNA

"DNA and the molecules that surround it form a truly superb mechanism - a miniaturised marvel. The information is so compactly stored that the amount of DNA necessary to code all the people living on our planet might fit into a space no larger than an aspirin tablet." **Paul S. Taylor,** - The Illustrated Origins Answer Book, page 23

"Life cannot have had a random beginning ... The trouble is that there are about two thousand enzymes, and the chance of obtaining them all in a random trial is only one part in 10 to the power of 40,000, an outrageously small probability that could not be faced even if the whole universe consisted of organic soup. If one is not prejudiced either by social beliefs or by a scientific training into the conviction that life originated on the Earth, this simple calculation wipes the idea entirely out of court..." **Fred Hoyle** & **Chandra Wickramasinghe,** - Evolution From Space

"The chance that useful DNA molecules would develop without a Designer are apparently zero. Then let me conclude by asking which came first - the DNA (which is essential for the synthesis of proteins) or the protein enzyme (DNA-polymerase) without which DNA synthesis is nil?...there is virtually no chance that chemical 'letters' would spontaneously produce coherent DNA and protein 'words.'" **George Howe,** - Expert In Biology Sciences

"...An intelligible communication via radio signal from some distant galaxy would be widely hailed as evidence of an intelligent source. Why then doesn't the message sequence on the DNA molecule also constitute prima facie evidence for an intelligent source? After all, DNA information is not just analogous to a message sequence such as Morse code, it is such a message sequence." **Charles B. Thaxton, Walter L. Bradley & Robert L. Olsen:** - The Mystery of Life's Origin, Reassessing Current Theories, New York Philosophical Library, 1984 pp 211-212

"Generation after generation, through countless cell divisions, the genetic heritage of living things is scrupulously preserved in DNA ... All of life depends on the accurate transmission of information. As genetic messages are passed through generations of dividing cells, even small mistakes can be life-threatening ... if mistakes were as rare as one in a million, 3000 mistakes would be made during each duplication of the human genome. Since the genome replicates about a million billion times in the course of building a human being from a single fertilized egg, it is unlikely that the human organism could tolerate such a high rate of error. In fact, the actual rate of mistakes is more like one in 10 billion." **Miroslav Radman & Robert Wagner,** - "The High Fidelity of DNA Duplication"... Scientific America. Vol. 299, No 2 August 1988, pp 40-44. Quote is from page 24

"In the meantime, the educated public continues to believe that Darwin has provided all the relevant answers by the magic formula of random mutations plus natural selection - quite unaware of the fact that random mutations turned out to be irrelevant and natural selection a tautology." **Arthur Koestler, -** Author

"Evolution lacks a scientifically acceptable explanation of the source of the precisely planned codes within cells without which there can be no specific proteins and hence, no life." **David A. Kaufman, -** Ph.D., University of Florida, Gainsesville

"Once we see, however, that the probability of life originating at random is so utterly minuscule as to make it absurd, it becomes sensible to think that the favourable properties of physics on which life depends are in every respect deliberate .... It is therefore almost inevitable that our own measure of intelligence must reflect ... higher intelligences ... even to the limit of God ... such a theory is so obvious that one wonders why it is not widely accepted as being self-evident. The reasons are psychological rather than scientific." **Sir Fred Hoyle, -** Well-Known British Mathematician, Astronomer & Cosmologist

"Ultimately, the Darwinian theory of Evolution is no more nor less than the great cosmogenic myth of the twentieth century." **Michael Denton, -** 'Evolution, A Theory in Crisis' Page 358

"Any suppression which undermines and destroys that very foundation on which scientific methodology and research was erected, Evolutionist or otherwise, cannot and must not be allowed to flourish ... It is a confrontation between scientific objectivity and ingrained prejudice - between logic and emotion - between fact and fiction ... In the final analysis, objective scientific logic has to prevail - no matter what the final result is - no matter how many time-honoured idols have to be discarded in the process ... After all, it is not the duty of science to defend the theory of Evolution and stick by it to the bitter end - no matter what illogical and unsupported conclusions it offers ... If in the process of impartial scientific logic, they find that creation by outside intelligence is the solution to our quandary, then let's cut the umbilical chord that tied us down to Darwin for such a long time. It is choking us and holding us back ... Every single concept advanced by the theory of Evolution (and amended thereafter) is imaginary as it is not supported by the scientifically established probability concepts. Darwin was wrong... The theory of Evolution may be the worst mistake made in science." **I. L. Cohen, -** Darwin Was Wrong - A Study in Probabilities P.O. Box 231, Greenvale, New York 11548: New Research Publications, Inc. pp 6-8, 209-210, 214-215. I.L.Cohen, Member of the New York Academy of Sciences and Officer of the Archaeological Institute of America.

"The notion that ... the operating programme of a living cell could be arrived at by chance in a primordial soup here on earth is evidently nonsense of a high order." **Sir Fred Hoyle, -** Evolutionist

"Far from being an established fact of science that it is so typically portrayed to be, Evolution is, in reality, an unreasonable and unfounded hypothesis that is riddled with countless scientific fallacies." **Scott M. Huse, -** The Collapse of Evolution, Baker Book House, Grand Rapids, Michigan, pp 127

"Unfortunately many scientists and non-scientists have made Evolution into a religion, something to be defended against infidels. In my experience, many students of biology - professors and textbook writers included - have been so carried away with the arguments for Evolution that they neglect to question it. They preach it ... College students, having gone through such a closed system of education, themselves become teachers, entering high schools to continue the process, using textbooks written by former classmates or professors. High standards of scholarship and teaching break down. Propaganda and the pursuit of power replace the pursuit knowledge. Education becomes a fraud." **George Kocan, -** "Evolution Isn't Faith But Theory", Chicago Tribune , p.9, Monday April 21 1980

"Scientists who go about teaching that Evolution is a fact of life are great con men, and the story they are telling may be the greatest hoax ever. In explaining Evolution we do not have one iota of fact." **Dr. T. N. Tahmisian,** - A Former U.S. Atomic Energy Commission Physiologist

"Evolution is a fairy tale for grown-ups. This theory has helped nothing in the progress of science. It is useless." **Dr. Louise Bounoure, -** Director of Research at the French National Centre for Scientific Research,

Director of the Zoological Museum and former president of the Biological Society of Strasbourg

Many Good Books

For additional reading:

Creation Trilogy, and What is Creation Science?, by Henry Morris, PhD. The late Dr. Morris was the founder of the Institute of Creation Research and the foremost thinker and author on Creation Science. All his books, and there are many, are brilliant. He has more science on one page than evolutionists have in an entire book. This trilogy would be ideal for a college course on creation science.

Refuting Evolution and Refuting Evolution 2 by Jonathan Sarfati. A simply written, but very scientific, rebuttal of the National Academy of Sciences official paper on creation and evolution. Intended for schools. Over 400,000 copies in print.

Dr. D. James Kennedy has about a dozen great books including Evolution and You and Evolution-The Root of the Problem. Dr. Kennedy is very good on this topic.

Darwin on Trial by Philip Johnson. Law professor conducts meticulous cross-examination of evolution to reveal total fraud. Recommended by the Discovery Institute.

Darwin's Black Box, by Michael Behe. Scholarly work on evolution. Recommended by the Discovery Institute.

The New Answers Book and War of World Views both edited by Ken Ham. Each is a collection of articles by different authors addressing various topics relating to evolution and creation.

War of World Views compiled by Gary Vaterlaus,

Collapse of Evolution by Scott Hughes,

Foundation Crumbling by Paul LaMoin,

Bone Peddlers by William Fix,

Evolution, a Theory in Crisis by Michael Denton

Two Good Organizations
Institute for Creation Research Museum and Research Center
10946 Woodside Ave.
North Santee, CA 92071
(214) 615-8300 (General Office)
(800) 628-7640 (Customer Service)

www.answersingenesis.org